T0319867

Rethinking Smart Cities

RETHINKING URBAN AND REGIONAL STUDIES

Books in this series enrich the study of cities and regions by promoting a cutting-edge approach to thought and analysis. Offering a forum for innovative scholarly writing, the series showcases authored books that address their subject from a new angle, expose the weaknesses of existing concepts and arguments, or 're-frame' the topic in some way. This might be through the introduction of radical ideas, through the integration of perspectives from other fields or even disciplines, through challenging existing paradigms, or simply through a level of analysis that elevates or sharpens our understanding of a subject.

Rethinking Smart Cities

Zaheer Allam

Chaire Entrepreneuriat Territoire Innovation, IAE Paris – Sorbonne Business School, Université Paris 1 Panthéon-Sorbonne, France

Yusra Raisah Takun

Live+Smart Research Laboratory, School of Architecture and Built Environment, Deakin University, Australia

RETHINKING URBAN AND REGIONAL STUDIES

Edward Elgar
PUBLISHING

Cheltenham, UK • Northampton, MA, USA

Published by
Edward Elgar Publishing Limited
The Lypiatts
15 Lansdown Road
Cheltenham
Glos GL50 2JA
UK

Edward Elgar Publishing, Inc.
William Pratt House
9 Dewey Court
Northampton
Massachusetts 01060
USA

A catalogue record for this book
is available from the British Library

Library of Congress Control Number: 2022946169

This book is available electronically in the **Elgar**online
Geography, Planning and Tourism subject collection
http://dx.doi.org/10.4337/9781803926803

ISBN 978 1 80392 679 7 (cased)
ISBN 978 1 80392 680 3 (eBook)

Printed and bound in Great Britain by TJ Books Limited, Padstow, Cornwall

Contents

About the authors

Zaheer Allam holds a PhD in Humanities, a Master of Arts (Res), an MBA and a Bachelor of Applied Science in Architectural Science from universities in Australia and the United Kingdom. Based in Mauritius, he is Chairperson of the National Youth Environment Council and a board member of the Mauritius Renewable Energy Agency. He works on a number of projects on the theme of smart cities and on strategies dealing with the increasing role of technology in culture and society. Zaheer is also the African Representative of the International Society of Biourbanism, a member of the Advisory Circle of the International Federation of Landscape Architects and a member of several other international bodies. He holds several awards and commendations and is the author of over 105 peer-reviewed publications and seven books about smart, sustainable and future cities.

Yusra Raisah Takun has a Master and a Bachelor's degree in Electronic, Communication and Computer Engineering. With extensive experience in large-scale architectural and urban projects in Mauritius, she has unique skills in design engineering and auditing and specialises in sustainable and smart building systems. Yusra also has an interest in emerging technologies, the pilot testing of new solutions and her research interests lie in the converging fields of architectural and urban governance with building and urban services.

Preface

Since the emergence of the COVID-19 pandemic over two years ago, the world has experienced unprecedented changes that could not have been anticipated, even after the 1918 Spanish flu. The widespread nature of COVID-19 has had far-reaching impacts in all global dimensions – economic, social, politics, environment, education and health. As a result, a majority of the global population has been left confused and uncertain about many things. This inspired us to write this book to show the world that despite the unprecedented challenges that our global cities have endured over the years, with COVID-19 compounding those issues, there is still hope to achieve resilience and sustainability as well as improve our urban liveability status. While lifestyle and human behaviours expected to change drastically post-pandemic, the objectives and targets of creating urban areas that have the capacity and potential to weather down issues of traffic, climate change, scarcity of resources, population increases and others is still on course.

The content of this book is further inspired by the realisation that despite the different urban concepts such as the compact city, sustainable city, creative city, smart city and sponge city, a wide constituent of the global population have little or no clue of such concepts. As a result, most of the people globally, as reported both in mainstream media and social media, wonder what will become of our cities, the nature of workplaces, our education systems and other areas that are pertinent to human existence. These fears are paramount despite a substantial majority of the global population having received at least one jab of the COVID-19 vaccination, and the severity of the disease seemingly having been reduced. Practically, the vaccination ought to have boosted the confidence levels of most people and allowed them to quickly adapt to the 'new normal', but there is enough evidence that this will take time. In fact, from a psychological point of view, it is feared that 'cave syndrome' might prevail in most places, especially in cities, hence jeopardising efforts to restore dimensions relating to society, politics and the economy.

The above background explains in part why this book was written, not only to share the history of our cities and why they remain the best bet for economic and social recovery, but also to highlight that the future of our cities remains bright. The hope for successful cities is pegged on a myriad of things, but this book emphasises the importance of existing and future technologies and their position in making cities more attractive for economic, social, environmental sustainability, educational and health reasons. In the course of advancing the need for different technologies to be embraced and applied in different urban endeavours, this book explores the contribution of already existing technology and how it has been instrumental, including in helping to solve the COVID-19 pandemic and issues that have come out of it. Further, it explores how these technologies have the potential to influence the emergence of new ones that will help take our cities to the next level where human-scale agendas will receive maximum attention.

Besides explaining the advantages of applying different technologies to cities, this book also explores some of the notable challenges that have the potential to derail this agenda and how they could be amicably addressed. The concern on the usage of data, in particular, has been addressed at length in this book, as we believe that it has huge potential to jeopardise the confidence that people have in technology, let alone their application in cities. Were it not for the extensive research that has been done in the different ways technology could be adopted to solve issues such as privacy and security concerns of data use in cities, there would have been serious pitfalls in those areas. However, we acknowledge that it is possible for data-driven cities to exist without major privacy or security threats being experienced. This is why we associate with numerous works done by researchers, academics, scientists and other stakeholders in the urban planning realm concerning the need for comprehensive public participation. As we show in this book, urban residents are an important constituent in the running and governance of urban areas, hence they need to be well informed, and where possible, actively involved in decision making, including in matters of technology application in cities and how these can be finance and managed.

In a bid to make clear some of the urban planning approaches that could be adopted to restore the confidence of most urban dwellers, we have dedicated substantial attention in exploring a viable emerging planning model – the 15-minute city concept. The concept is hinged on the promise of promoting the human dimension in our cities by ensuring that urban residents have access to basic amenities within their neighbourhoods in less than 15 minutes by either walking or cycling. This sounds utopian, but it could be achieved, especially with the determination to reduce automobile use as well as promote social distancing that came about during the height of COVID-19. We highlight how this concept will become even more viable when coupled with anticipated advanced technologies such as 6G, comprehensive artificial intelligence, immersive reality and digital twins. This concept will not only help achieve the human dimension, but expectations are that it will also help achieve other global objectives like the decarbonisation agenda that was passionately advocated during COP 26.

While this book supports the deployment of technology in the pursuit of resilience, adaptability and liveability in cities, it does not promote technology as the silver bullet for all major issues. It has therefore been acknowledged that there are other elements that need to be collated together to make cities – now and in the future – more vibrant in all aspects, including the economy, society, politics and environmental sustainability.

1. What is a smart city? Understanding the concept beyond a tech-centric approach

INTRODUCTION

The history of cities dates back to the agrarian revolution, where due to an increase in agricultural productivity, people started to be attached to specific locations near food production areas, the main aim being to increase liveability, reduce transportation cost and allow for storage of surplus products. As a result, those locations became centres for trade and settlement for people offering different goods and services. During this period, technologies were rudimentary and not as advanced as they are today, but still prompted people to pool together in specific locations, which eventually grew to become towns and cities (Maisels, 1993). After the emergence of the Industrial Revolution, the role of cities continued to be more prominent, where they became centres for economic development, marketplaces for diverse products, homes for factories and manufacturing and others. This prompted an increase in opportunities – economic, social and tourism activities – and attracted a large influx of people (Allam, 2020c). As a result, existing towns transformed into cities, while cities grew and became even larger. But, just like in the case of the cities during the agrarian revolution, the common denominator was the presence of technology.

Currently, the world is experiencing an unprecedented growth in advanced technology following the emergence of the fourth industrial revolution. Characterised by massive use of information technology (IT), cities have continued to become the melting pot for numerous economic, social and political activities (Allam, 2020e, 2021c). They have become centres for higher education, macro-economic activities, health, tourism destinations, entertainment centres and home to multinational companies and organisations (Colenbrander, 2016). They have also become home to over 55 per cent of the global population and it is expected that by 2050 between 68 and 70 per cent of the global population will be living in urbanised areas (UN Habitat, 2020). This substantially large population will be prompted by a myriad of factors, including rapid urbanisation and a reduction in economically viable activities in rural areas; hence forcing people, especially the younger generation, to seek opportunities in cities (Yu et al., 2019). Further, the growth and subsequent use of advanced technologies is seen to prompt a change and emergence of new economic activities, aided by technology, which is attracting more people into urban areas.

The use of IT products in cities is not only based on economic and social benefits, but it is also being openly deployed to offer solutions to increasing urban challenges that modern cities are facing (Allam, 2021b). Among such include the ranging impacts of climate change which have prompted unprecedented challenges, especially in coastal cities, small island developing states (SIDS) (UNEP, 2014) and least developed countries (LDCs) (UNCTAD, 2020). But the impacts are also notable even in developed economies and cities not within coastal or low-lying regions (Costa & Floater, 2015; Forzieri et al., 2018; Kulp & Strauss, 2019). On this, it is reported that over 100 cities across the globe are at high risk of serious climate change incidences like increased and erratic precipitation, heatwaves, flooding and others. In some cases, it is reported that a sizeable number of SIDS are facing the danger of being submerged, despite their contribution to the global greenhouse gas emissions being almost negligible (less than 1 per cent) (United Nations Department of Economic and Social Affairs, 2019). But, while cities face the greatest risk, it is appreciated that they are also – paradoxically in the forefront of contributions to climate change. As of 2020, cities were producing approximately 60 per cent of global emissions, especially due to high demand in the energy and transport sectors (United Nations, 2020). Therefore, the high adoption of technologies, especially to enhance urban planning and urban regeneration, is informed by how these could help in the search for solutions for some of the aforementioned challenges.

In addition to climate change, cities across the globe are experiencing notable challenges in the transport sector. However, diverse smart technologies have strong capacities to mitigate most of these issues (Badassa et al., 2020; Gambella et al., 2019). Such challenges range from traffic congestion, which for instance is reported to prompt a loss of more than $88 billion in the United States due to traffic delays per year (Rumsey, 2022). This also led to a total loss of more than $74.1 billion in the freight industry, with cities experiencing over $66.1 billion in 2019

(McCarthy, 2020). Congestion also prompts an equivalent of eight days in the number of working hours lost, and over 41 per cent of emissions released in the environment attributable to the transport sector (Fleming, 2019). Besides the direct cost, the transport sector has been attributed to conventional planning models, characterised in most cities across the globe. On this, it has been argued that most cities are fashioned to allow for smooth vehicular flows, hence resulting in gridlocked linear streets that are not conducive for social dimensions. Such planning models have faced criticism from urbanists like Salingaros (2000), Alexander (1979, 2002), Jacobs (1961) and others, who find that most cities are not well structured for human interactions, human recreation purposes or human health endeavours such as walking or cycling. On the contrary, most are seen to cherish the glamour of the built environment; high-rise buildings and mega transportation architectural projects. These put little value into walkable streets, multiuse green spaces or public spaces that would facilitate active social interactions.

With the need for urban regeneration, different existing and emerging technologies have been tried and tested in varying geographies across the globe, and there is evidence of positive results. For instance, smart technologies such as artificial intelligence (AI), the Internet of Things (IoT), big data, cloud computing and others have been deployed in cities like Barcelona (Gascó-Hernandez, 2018), Songdo in South Korea (Kolotouchkina & Seisdedos, 2017), Singapore (Rohaidi, 2018) and others and sectors like transport, health, education and the environment have been seen to be positively transformed (Allam & Dhunny, 2019; Allam & Jones, 2018). For instance, in cities like London, Singapore and others, it is reported that public service delivery has improved due to the unification of service delivery platforms, with citizens able to engage with local government in real time. In Singapore, it is reported that energy consumption is reported to have greatly reduced after the government formulated smart technology policies that targeted the housing sector (Bhati et al., 2017).

It is important to note, however, that most smart technologies are still in their infancy stages and require much testing, standardisation of protocols and communication frameworks (Allam & Dhunny, 2019; Allam & Newman, 2018). The concerted efforts and much attention that the smart city is receiving from both the public and private sectors will lead to acceptable implementation pathways. With this backdrop, this chapter explores how smart cities emerged and how technology plays a critical role in its actualisation.

TECHNOLOGY: THE ANCHOR TO MODERN URBAN PLANNING

The history of cities is amass with numerous transformational changes that have been effected with the objectives of bringing in notable changes. For instance, they make cities more efficient in resource consumption, make them attractive for social, economic and political activities and make attraction sites for visitors (Allam & Newman, 2018). The outcome of these transformations is an increase in the urban population as well as rapid urbanisation experienced especially during this century as captured in a report by UN Habitat (2020). However, the most influential transformation in most cities can be attributed to the era when technology adoption became part of urban planning. Previously, urban planning and designs were shaped by the emergence of automobiles (Selzer & Lanzendorf, 2019), and as a result, it was important for cities to be ordered in a way that could accommodate central business districts, sprawling suburbs and big ring roads (Newman & Kenworthy, 2015). But as people started to increase in urban areas coupled with unprecedented lifestyle changes (the way people work, shop, enjoy recreation and entertainment and travel, etc.), the auto-centric planning models became untenable. With the need to 'repair' the physical and social fragments that contemporary planning brought about, there is an ongoing quest to equip cities and different urban fabrics with advanced technologies that support lifestyle changes, increasing urban challenges and future planning prospects (Allam, 2020g, 2021c).

At the core of urban planning are diverse technologies ranging from Industry 4.0, IoT, AI, big data and data computation technologies that make it possible for planners, designers and other stakeholders to integrate the 'smartness' aspects in cities. Industry 4.0 technology is among the key technologies that have allowed for the integration of different smart components. These include the smart environment, smart living, smart mobility, smart economy, smart population, intelligent infrastructures and others into cities' main fabric (Allam, 2021a; Safiullin et al., 2019). This technology has made it possible to advance other technologies such as AI and IoT and to make them part of the smart urban system. In particular, these are very critical in the management of increasing urban data being collected and subsequently used to inform decisions on different urban aspects. IoT is critical in supporting the manufacturing and integration of different smart products (things) that serve as the basis for data generation. The intelligence of these products, however, is enabled by the availability of AI technology, which allows them not only to communicate with each other, but is also instrumental in the real-time collection and transfer of data to main central networks or multihoming. Multihoming is a network that integrates diverse networks within a system into a single network or environment (Rathee et al., 2021). With AI, smart things are gaining artificial capabilities to 'reason', learn, discover meaning and generalise past experiences. Coupled with other technologies such as big data, AI is argued to provide IoT devices with the capacity to allow for predictive and prescriptive analyses that are key in decision making (Allam & Dhunny, 2019).

Another contribution of different technologies in the achievement of the smart city concept is enabling the automation of different processes; hence, reducing human interventions and human-induced decision making. Advanced technologies such as artificial neural networks when integrated into IoT and cloud-based computing technologies have made it possible for urban processes to run autonomously, thus increasing the speed at which the analysis and processing of information and generation of insights are made (Allam, 2019a; Huang, 2017; Perwej & Parwej, 2012). Such automation has not only reduced labour-related costs, but is argued to help reduce security risks and propagate a culture of efficiency in the monitoring and processing of insights (Allam, 2019a). A case in point is how AI-based technologies have been incorporated into the transport system in Singapore, where law enforcement agencies are able to automatically monitor and force implementation of things such as tax returns by car users, adherence to speed limits, usage of parking slots and many other related services (Allam & Jones, 2018). While these are seen to be intrusive, dubbed as being akin to 'big brother', they have helped the island country to increase efficiency and security on roads (GovTech Singapore, 2017). The same conclusion on efficiency can be drawn from how smart technologies have been integrated in the entertainment sector, especially in sports events like soccer in most urban-based leagues across the world (Prensa, 2019), hence cementing the role of cities as the ultimate attraction centre for most people. On this, the introduction of video assistant referee (VAR), which is integrated with AI technology, helps in supporting human decision making in soccer, thus reducing 'costly' outcomes warranted by errors in human judgement (Samuel et al., 2020; Spitz et al., 2021).

In addressing endemic urban challenges like climate change, uncontrolled resource consumption, insufficient energy generation and distribution and traffic congestion, smart technologies have played a critical role. For instance, it is reported that different climate change events are causing real danger in most cities, especially those located in coastal regions and low-lying geographical areas and most SIDS (UNEP, 2014). It is not surprising that most cities in those regions are now experiencing unprecedented erratic precipitation, flooding, sporadic heatwaves and many other related climate change challenges (UNFCCC, n.d.; United Nations Department of Economic and Social Affairs, 2019). In an effort to mitigate and reduce the impacts of climate change, technologies such as AI, IoT and others play a critical role especially in helping to make future predictions on issues like frequency, velocity and magnitude of the climate events anticipated. With data generated from different sources, learning algorithms and sensing devices, it has become possible for relevant agencies to make informed decisions, hence enhancing their risk management approaches.

The prospect of even more technologies such as 5G and 6G mobile communication technology, the digital twins, advanced augmented reality, biomimicry especially in data storage (DNA storage) and others are expected to enhance the actualisation of the smart city concept further (Allam, 2020c). These in particular will have significant impacts on data generation, storage, transfer and manipulation, which are critical for making cities experience even more smart services.

THE SMART CITY CONCEPT

The quest to transform urban areas by way of technology has been in existence since the 1970s, with the city of Los Angeles being the pioneer. Through its Community Analysis Bureau, the city of Los Angeles relied on data and aerial photograph analysis for different urban planning purposes (Soja et al., 1983). In particular, information retrieved by the organisation facilitated resource allocation, report formulation, disaster and poverty mitigation among other things with more effectiveness (Brasuell, 2015). However, during this period, technology was rudimentary, hence making it harder for the Los Angeles model to be duplicated in other cities. Later, as the fourth industrial revolution emerged (in the 2000s), different technologies started to be developed, allowing technology service providers to collaborate with municipalities to use data generated and collected from different urban fabrics, finding solutions for diverse urban challenges (Allam, 2020d). This was for instance witnessed in Amsterdam, where the first virtual digital city is argued to have been created (Alberts et al., 2017). However, it was the work of IBM and Cisco in the mid-2000s that brought real breakthroughs in the use of technology in cities to render them 'smart' (Allam & Newman, 2018). In 2011, the first Smart City Expo World Congress was inaugurated in Barcelona and since then much progress has been achieved, with the Expo becoming an annual activity (Gascó-Hernandez, 2018).

The initial smart city programmes started with transformation in key sectors such as the energy sector where smart meters and smart grids became popular (Dedrick & Zheng, 2011). The first smart grid project was initiated in the United States in 2009 through the American Recovery and Reinvestment Act, which allowed for its funding. In Europe, the smart solution in the energy sector was adopted in the form of the Smart Meters Project, initiated in 2009 with a target of reaching approximately 80 per cent of consumers by 2020. These were followed in 2011 by the city of Barcelona which deployed data-driven street lighting, an urban smart upgrade system that also included public transit and parking (Gutiérrez Escolar et al., 2015). Overall, since the inception of 'smart' energy dimensions, notable achievements have been realised including an emphasis on renewable energy even in energy-intensive countries like China (IEA, 2017) and in oil-rich countries like Kuwait (KUNA, 2018).

Notable transformations have also been witnessed in the transport sector across different cities, with projects such as the installation of sensors, cameras, street lighting and others gaining traction. There has also been an increase in the manufacturing and production of electric vehicles and autonomous vehicles, the emergence of ride sharing and other smart startup projects, all making the transport sector more sustainable for cities (McKinsey & Company, 2018). The backbone of the different smart transformations in the transport sector, mainly targeting sustainability, are due to the increase in the amount of data being shared from a myriad of sources (Allam, 2020d). In particular, the most reliable sources of data have been the different sensors and cameras deployed in transport infrastructures to facilitate different agencies, especially law enforcement agencies in managing and controlling traffic, parking, use of cycling lanes, footpaths and others. Data are also being deployed to facilitate the adoption of 'smarter' options, such as the multiuse of parking spaces for activities such as recreation (Anuar, 2017).

The diverse applicability of the smart city concept in shaping different urban fabrics and dimensions has seen it embraced across the globe, including in LDCs, despite the capital-intensive nature of these projects (Azharianfar & Kermani, 2016). This is firmly demonstrated by the increasing number of smart cities and the activities thereof. For instance, by the end of 2020, growing at an approximated compounded annual growth rate of 18.22 per cent, the smart cities market reached a high of about $739.78 billion and is expected to break the ceiling and grow to about $2.04 trillion (Mordor Intelligence, 2021) or $3 trillion (Smart Cities Association, 2021) by 2026 as projected by different companies. The acceptability of the smart city concept is accentuated by the emergence and growth of supportive technologies and smart devices. For example, IoT, which is part of the critical component of smart cities, has been growing at an exponential rate since its emergence. By 2015, when the smart city concept started to gain more prominence, there were only 3.6 billion connected IoT devices (Liu, 2019). However, the growth has been astronomic, as connected devices increased to approximately 13.8 billion by 2021 and are expected to reach a high of between 30.9 billion (Vailshery, 2021) and 75 billion (Liu, 2019) by 2025.

The growth of the smart city and IoT markets has had significant impacts on the economy, society, administration and environment in cities where they have been or continue to be implemented. On the economic front, there is evidence of the emergence of new startups, which together with existing economic frontiers have prompted numerous economic benefits. For instance, it is reported that in 2014 alone, Barcelona managed to save over 75 million euros by automating part of its public service (Kamel Boulos et al., 2015). In the transport sector, adopting 'smart' technologies is argued to have the capacity to help cities save costs on congestion, emissions and others equivalent to $800 billion per annum (Nastu, 2020). The smart city concept further helps bolster employment opportunities and enhance efficiency and performance (Bibri, 2020); factors that have always had negative impacts on most conventional cities.

Socially, smart cities have been associated with enhancing liveability status, following a reduction in emissions in sectors like transport, housing, waste management, manufacturing and others. This is further promoted by the adoption of safe and more sustainable energy generation options, like the case of Fujisawa in Japan, which has all of its 1,000 homes powered by solar energy (Inhabitat, 2011). As a result, besides having reliable energy, the city has also managed to reduce approximately 70 per cent of its carbon emissions. The social improvement associated with smart cities is also highlighted by issues like reduced health challenges, improved income status of urban residents and the availability and distribution of basic services such as water, waste management and others efficiently and on demand.

Despite the notable benefits associated with the smart cities concept, there are some challenges that still require ironing out. One identified challenge is that of financing. On this, it has been correctly argued that only a few cities across the globe have the financial capacity to fully finance the capital-intensive urban planning model. In most cases, a majority of the cities that have some elements of 'smartness' have managed to do so by engaging the private sector (including via public-private partnership schemes), or securing funding from diverse financial institutions (Fishman & Flynn, 2018). The challenge in particular has been amplified in cities in the global south, where already the challenge of the external debt burden is reported as a serious challenge not only for cities but also for many national governments. Cities in those economies find it extremely difficult to secure funding for projects, including critical ones like infrastructure development (Deloitte, 2013). In addition to funding, the implementation of the smart city concept further faces challenges emanating from competition by smart service and smart product providers, who in most cases are motivated by the prospect of increasing their profits. As a result, challenges of data management, connectivity, standardisation of protocols and other elements continue to abound as each provider seeks to gain a competitive advantage and control over the lucrative smart city market (Allam, 2020a). However, despite the presence of those challenges, the potential of this planning concept in transforming cities to withstand the myriad urban challenges underscores its value and the hope of it becoming even more popular in the future. Further, with continued activity in the technology frontier, where the advancement and emergence of new technologies is becoming more prominent, the pursuit of smart cities will hold. A case in point is how smart technologies mitigated the economic crisis during the COVID-19 pandemic, as some critical economic activities, especially in the service sector, continued, albeit at a smaller scale (Allam, 2020h).

'SMARTNESS' IS NOT ALL THAT CITIES NEED

The prospect of making cities 'smart' has been gaining traction in different parts of the world, especially in most developed and developing economies, and it is anticipated that it will become common even in LDCs. Its prospects are majorly highlighted by how the urban planning model is helping to address diverse challenges and negatives associated with modern cities. Accordingly, substantial attention to the model is associated with the potential to help urban planners, designers and managers incorporate aspects that resonate with the philosophical undertones of urbanists such as Christopher Alexander (1965, 2002; Alexander et al., 1977), Jane Jacobs (1961), Jan Gehl (2013) and others who have been advocating for cities that not only value material things but that are also built for people. With different smart technologies, it is becoming possible to build smart cities from scratch (e.g. Songdo in South Korea) as well as integrating those technologies in existing urban infrastructures (e.g. Barcelona, London, New York, Singapore, Amsterdam) (IDC, 2018). Either way, conventional city planning models are being replaced by the smart city model and, in the future, cities that are digitally inclined, and technologically advanced cities, will be increasingly popular (Allam, 2020e; Allam & Jones, 2021).

The aspect of 'smartness' alone, however, cannot suffice to transform and address the numerous issues associated with current cities. In part, one would be technically correct to argue that the impacts of the numerous smart technologies being deployed for smart city programmes have the capacity to transcend and address other aspects such as safety and social factors (Allam, 2019; Rodriguez-Segura, 2020; UNCTAD, 2021). However, it has been acknowledged that while that is possible, numerous underlying factors have made it complicated for the smart city concept to address all those dimensions. Factors like the competition that exists between smart service and device providers, their profit-centric structuring and the lack of standardisation of protocols and frameworks that would allow for seamless communication and sharing of platforms and networks among other things are just a few inhibitors (Allam, 2019, 2019c, 2020f). In addition, the handling of data, the quality thereof and the apprehension of urban residents to share private and personal information further affirm the perception of critics, especially in regard to privacy and security (Ismagilova et al., 2020). In this case, it is paramount for urban managers to extend their attention and effort and deploy other approaches to ensure that cities are also safe, resilient, adaptable, liveable, economically feasible and sustainable.

The smart city concept could be better structured and managed to ensure that all the aforementioned attributes are incorporated. This could be achieved via the deployment of multipronged approaches that ensure that there is cordial cooperation and collaboration between different interested parties and agencies during the lifecycle of the project (Ristvej et al., 2020). The argument is that the smart city concept should form the basis for other attributes to be achieved in, and within, the city, without the need for introducing a parallel set of strategies or technologies. The aspect of safety, for instance, is crucial in determining the liveability status of a city: the higher the safety index, the higher the liveability status is. It is therefore paramount that this aspect is emphasised, as is the case with 'smartness'. On this, safety could be viewed not only in terms of a reduction in insecurity resulting from crime and delinquency, but should be further extended to address the perceived fear of loss of private data and compromise in personal privacy (Ismagilova et al., 2020; Zoonen, 2016). Such fears, unfortunately, have been observed to be accentuated by the availability of some smart digital solutions, such as the use of unmanned aerial vehicles like drones, which critics argue have the potential to infringe on the privacy of residents (Mohamed et al., 2020). Similarly, the use of cameras and other image-capturing devices across different urban installations can discourage people from participating in urban activities, or hinder the quality of urban open spaces (Zoonen, 2016). But, where such are installed with the objective of not only providing smartness to a city but also promoting safety, positive outcomes have been achieved (Ristvej et al., 2020). A case in point is the city of Hangzhou in China (Argyriou, 2019), which is considered to be among the safest cities in terms of traffic following a robust deployment of smart technologies, dubbed the 'City Brain' (Argyriou, 2019).

In his theory on urban complexity and coherence, Salingaros (2000) emphasised the need for different large-scale wholes to be assembled and ordered in a manner that they ultimately lead to a 'whole' structure. In those 'wholes', he insisted that small-scale elements and components are equally important as they eventually add up to form a large-scale coherence. This theory, however, does not seem to be applied in most conventional cities, perhaps explaining why different urban elements appear to be disjointed from one another, causing numerous problems. For instance, the transport sector, which in theory should allow for smooth and quick accessibility of different urban nodes, has unfortunately been the source of a sizeable number of urban challenges. The increasing number of private cars, uncontrolled use of non-renewable energy, disruption of other mobility options such as cycling and walking and gridlocked planning approaches are just a few notable issues that make it hard for cities to be coherently ordered. With the smart city concept, the intention is that such challenges can be addressed by increasing efficiency and performance, as well as introducing new and alternative ways of addressing emissions, parking issues and reducing accidents (Allam &

Newman, 2018). But, these have been overshadowed by the egocentric nature of most corporations providing data services or smart products. In the case of urban regeneration accommodating the 'smartness' perspective, it has been observed that, in addition to cities becoming 'smart', this leads to gentrification; thus, the poor and other vulnerable urban dwellers are forced to move out of their traditional areas (Allam, 2019b, 2020i). Such events happen due to factors like price increases for urban properties, the cost of living and in accessing diverse basic services. This makes the areas become too expensive for such groups to live in, and as Jane Jacobs (1961) reiterated, these groups then become alienated.

The negative aspects associated with the pursuit of urban 'smartness', however, need to be overcome as they have the potential to degenerate expensive endeavours that might eventually lead to ghost projects (Malanga, 2018). This would plunge not only the economy of cities into crisis, but also that of entire countries in some cases, as most smart city programmes being undertaken in different regions globally are credit financed (Andersen, 2018). Therefore, the return on investment for such projects needs to be guaranteed, otherwise it could lead to endangering the stability of an entire economy (Allam, 2019).

SMART CITY HURDLES AND SOLUTIONS

Smart city programmes are viable for both existing and emerging urban areas, with most of the activities of incorporating smart technologies seen to be taking place in existing cities. This is understandable, as such cities have existing infrastructure, processes and activities ongoing that are only required to be updated and restructured to align with smart programmes (infrastructural modernisation) (Allam & Society, 2020). The existence of infrastructure is usually argued to be one of the challenges that the smart city concept has to overcome, as most of them are not fashioned to accommodate smart solutions and would require being restructured or changed (Allam, 2012, 2017, 2019d). Most existing infrastructures were rendered to offer and allow for services other than 'smart' services. After all, the smart city concept came into light and started gaining traction in the 2000s after the fourth industrial revolution emerged (McKinsey & Company, 2018), with the concept peaking in 2015 (Allam & Newman, 2018). Therefore, to ensure that the infrastructures have the capacity to support smart city programmes, there is a need for regeneration, especially regarding energy, to ensure the smart devices are sufficiently powered and are able to provide back-up energy in case of blackouts that would render many smart devices inaccessible (Nižetić et al., 2020). Further, such regeneration programmes allow for increased energy demand prompted by smart devices to be met without compromising the energy distribution in other urban sectors (Porphyrios et al., 1985). In addition to energy, smart city infrastructures prompt further demand for security installations to guarantee that they can always run without physical damage or compromises (Yao et al., 2021).

In the course of such regeneration, identifiable trends include the total replacement of some of the existing infrastructure, alteration of urban architectural plans and temporary disruption of normalcy during the building, maintenance and repairing of infrastructure. Such activities may have notable implications on urban activities. In addition, the most obvious challenge is the increase in the financial burden to cities. Infrastructural investments are relatively expensive and demanding, and the integration of smart solutions only adds to infrastructural development burdens (Oughton et al., 2019; Sharma et al., 2020). The gist of the matter is that most cities rely on external sources for their investments and this has the potential to plunge cities into unprecedented debt, thus derailing smooth service delivery (Deloitte, 2018). Also, demands by external financiers have the potential to cause some cities to shelve their programmes as they may not have the capacity to meet the terms and conditions for debt advances. Financing demands for smart city projects also seem to be competing with other services and this is difficult to balance, especially in cases where the general public is insufficiently educated, informed and engaged in the ongoing programmes. This exposes smart projects to the risk of not being embraced by local residents, which might lead to apprehension in the sharing of critical data among other challenges that might cause the programme to be successful (McKinsey & Company, 2018).

The smart city concept is hailed for its potential to improve the quality of life of urban residents through improved governance, the adoption of smart housing, improved security and the availability of real-time solutions to different urban issues. These are influenced by the amount and quality of data being corrected and analysed and the decision derived thereof. However, in the course of data correction, concerns on the need for balance between 'smartness' and privacy keep arising (Allam, 2019c, 2020f). In such cases, urban residents always favour their privacy (Martinez-Balleste et al., 2013). The aspect of 'Big Brother', where residents feel like they are being monitored, is abhorred, and for law-abiding citizens, is seen to instil fear and paranoia (De Guimarães et al., 2020).

Privacy concerns arise when local governments and those involved in providing smart solutions barely involve local residents, hence arousing scepticism on where the data are sourced and their intended purpose. In other scenarios, such data have been reported to be exposed to third parties, especially by the profit-oriented tech companies contracted to offer smart services. The fear here is associated with the potential for cyberattacks as most third parties may not be keen to deploy extra security measures to safeguard (including encrypting) the data (El Emam et al., 2015).

The security threats in most cases are prompted by potential vulnerabilities in products that different companies offer, as their concentration is more on increasing the number of devices in the market to outperform their competitors (Woetzel et al., 2018). Most of the privacy concerns could be addressed by actively engaging local residents in different stages of smart city project implementation, more so during the initial stages where they are educated and informed of the purpose for data collection, the technology to be used and the expected benefits and negative implications if any (Calzada, 2018). In the case of IoT devices and technologies, the vulnerability questions could be addressed by relevant authorities devising policies and guidelines to be followed before products are deployed in the market. Further, standardisation of protocols and communication networks needs to be emphasised to ensure that all the installed technologies and devices can be monitored via a manageable number of networks (Rathee et al., 2021). After making sure device communication is congruent with agreeable standards, data storage challenges also need to be addressed; these have been a major concern in most smart cities. This challenge is often prompted by an unprecedented increase in the amount of data being generated and stored. Different options could be adopted for storage purposes, including exploring the use of biological options like the use of DNA and protein as storage alternatives. However, the gradual adoption of 5G technology and the prospect of 6G technology in the future are expected to allow huge data to be transferred in real time, hence helping to mitigate the storage challenges as they will help address the long buffer times, delays between uploading and sharing of content (Allam & Jones, 2021; Peters & Besley, 2019).

Another major hurdle that the smart city concept needs to overcome relates to inclusivity, both socially and economically. Most smart city solutions have the potential to alienate some demographics, hence preventing them from experiencing the benefits. For instance, in the mobility sector, it has been identified that most smart solutions are accessible via smart devices, especially those with the capacity to use mobile apps. While there is over 78 per cent of mobile phone penetration globally (with some cities recording over 90 per cent), some demographics like the elderly, the poor and people with physical disabilities may not have the luxury of using such devices or have knowledge of apps. This means that they can be left out in accessing the smart services. The same applies in areas like smart housing, smart parking and the adoption of small-scale alternative energy like rooftop solar power. This leads to the need to first address the digital divide, prior to engaging in smart city projects, to ensure that the product is beneficial to all, irrespective of social status, wealth or location.

REFERENCES

Alberts, G., Went, M., & Jansma, R. (2017). Archaeology of the Amsterdam digital city: Why digital data are dynamic and should be treated accordingly. *Internet Histories*, *1*(1–2), 146–159.

Alexander, C. (1965). A city is not a tree. *Architectural Forum*, *122*(1), 58–61.

Alexander, C. (1979). *The Timeless Way of Building*. Oxford University Press.

Alexander, C. (2002). *The Nature of Order: The Process of Creating Life*. Centre for Environmental Structure.

Alexander, C., Ishikawa, S., & Silverstein, M. (1977). *A Pattern Language*. Oxford University Press.

Allam, M. Z. (2019). *Urban Resilience and Economic Equity in an Era of Global Climate Crisis*. University of Sydney.

Allam, Z. (2012). Sustainable architecture: Utopia or feasible reality? *Journal of Biourbanism*, *2*(1), 47–61.

Allam, Z. (2017). Building a conceptual framework for smarting an existing city in Mauritius. *Journal of Biourbanism*, *17*(1–2), 103–121.

Allam, Z. (2019a). Achieving neuroplasticity in artificial neural networks through smart cities. *Smart Cities*, *2*(2), 118–134.

Allam, Z. (2019b). The city of the living or the dead: On the ethics and morality of land use for graveyards in a rapidly urbanised world. *Land Use Policy*, *87*, 104037.

Allam, Z. (2019c). The emergence of anti-privacy and control at the nexus between the concepts of safe city and smart city. *Smart Cities*, *2*(1), 96–105.

Allam, Z. (2019d). Identified priorities for smart urban regeneration: Focus group findings from the city of Port Louis, Mauritius. *Journal of Urban Regeneration Renewal*, *12*(4), 376–389.

Allam, Z. (2020a). Actualizing big data through revised data protocols to render more accurate infectious disease monitoring and modeling. *Surveying the COVID-19 Pandemic and Its Implications*, 71.

Allam, Z. (2020b). *Cities and the Digital Revolution: Aligning Technology and Humanity*. Springer International Publishing.

Allam, Z. (2020c). Data as the new driving gears of urbanization. In *Cities and the Digital Revolution: Aligning Technology and Humanity* (pp. 1–29). Springer International Publishing.

Allam, Z. (2020d). On culture, technology and global cities. In *Cities and the Digital Revolution: Aligning Technology and Humanity* (pp. 107–124). Springer International Publishing.

Allam, Z. (2020e). Privatization and privacy in the digital city. In *Cities and the Digital Revolution: Aligning Technology and Humanity* (pp. 85–106). Springer International Publishing.

Allam, Z. (2020f). *The Rise of Autonomous Smart Cities*. Springer International Publishing.

Allam, Z. (2020g). *Surveying the COVID-19 Pandemic and Its Implications: Urban Health, Data Technology and Political Economy*. Elsevier Science.

Allam, Z. (2020h). Urban and graveyard sprawl: The unsustainability of death. In *Theology and Urban Sustainability* (pp. 37–52). Springer International Publishing.

Allam, Z. (2021a). Big data, artificial intelligence and the rise of autonomous smart cities. In *The Rise of Autonomous Smart Cities* (pp. 7–30). Springer International Publishing.

Allam, Z. (2021b). The case for autonomous smart cities in the wake of climate change. In *The Rise of Autonomous Smart Cities* (pp. 61–74). Springer International Publishing.

Allam, Z. (2021c). On complexity, connectivity and autonomy in future cities. In *The Rise of Autonomous Smart Cities* (pp. 31–47). Springer International Publishing.

Allam, Z., & Dhunny, Z. A. (2019). On big data, artificial intelligence and smart cities. *Cities, 89*, 80–91.

Allam, Z., & Jones, D. (2018). *Promoting resilience, liveability and sustainability through landscape architectural design: A conceptual framework for Port Louis, Mauritius, a Small Island Developing State*. IFLA World Congress Singapore.

Allam, Z., & Jones, D. (2021). Future (post-COVID) digital, smart and sustainable cities in the wake of 6G: Digital twins, immersive realities and new urban economies. *Land Use Policy, 101*, 105201.

Allam, Z., & Newman, P. (2018). Redefining the smart city: Culture, metabolism and governance. *Smart Cities, 1*(1), 4–25.

Allam, Z., & Society. (2020). Sustainability and resilience in megacities through energy diversification, land fragmentation and fiscal mechanisms. *Sustainable Cities, 53*, 101841.

Andersen, C. L. (2018). Balancing the act: Managing the public purse. *Finance and Development, 55*. www.imf.org/external/pubs/ft/fandd/2018/03/pdf/fd0318.pdf

Anuar, A. N. (2017). The demand of recreational facilities in neighborhood parks. *Planning Malaysia, 16*(7).

Argyriou, I. (2019). The smart city of Hangzhou, China: The case of Dream Town Internet village. In L. Anthopoulos (Ed.), *Smart City Emergence* (pp. 195–218). Elsevier.

Azharianfar, S., & Kermani, A. (2016). Implementation of smart cities in the developing countries. 33rd International Geographical Congress, Beijing.

Badassa, B. B., Sun, B., & Qiao, L. (2020). Sustainable transport infrastructure and economic returns: A bibliometric and visualization analysis. *Sustainability, 12*(5). https://doi.org/10.3390/su12052033

Bhati, A., Hansen, M., & Chan, C. M. (2017). Energy conservation through smart homes in a smart city: A lesson for Singapore households. *Energy Policy, 104,* 230–239.

Bibri, S. E. (2020). Data-driven environmental solutions for smart sustainable cities: Strategies and pathways for energy efficiency and pollution reduction. *Euro-Mediterranean Journal for Environmental Integration, 5*(3), 66.

Brasuell, J. (2015, 22 June). The Early History of the 'Smart Cities' Movement – in 1974 Los Angeles. Planetizen. Retrieved 20 January 2021 from https://www.planetizen.com/node/78847

Calzada, I. (2018). (Smart) citizens: From data providers to decision-makers? The case study of Barcelona. *Sustainability, 10,* 3252.

Colenbrander, S. (2016). *Cities as Engines of Economic Growth: The Case for Providing Basic Infrastructure and Services in Urban Areas.* IIED.

Costa, H., & Floater, G. (2015). Economic costs of heat and flooding in cities: Cost and economic data for the European Clearinghouse databases. *Reconciling Adaptation, Mitigation and Sustainable Development of Cities.* https://climate-adapt.eea.europa.eu/metadata/publications/economic-costs-of-heat-and-flooding-in-cities/ramses_2015_economic-costs-of-climate-change-in-european-cities.pdf

De Guimarães, J. C. F., Severo, E. A., Felix Júnior, L. A., Da Costa, W. P. L. B., & Salmoria, F. T. (2020). Governance and quality of life in smart cities: Towards sustainable development goals. *Journal of Cleaner Production, 253,* 119926.

Dedrick, J., & Zheng, Y. (2011). Smart grid adoption: A strategic institutional perspective. Industry Study Conference.

Deloitte. (2013). Funding options: Alternative financing for infrastructure development. Retrieved from www2.deloitte.com/content/dam/Deloitte/au/Documents/public-sector/deloitte-au-ps-funding-options-alternative-financing-infrastructure-development-170914.pdf

Deloitte. (2018). The challenge of paying for smart cities projects. Retrieved 23 January 2022 from https://www2.deloitte.com/content/dam/Deloitte/us/Documents/public-sector/us-ps-the-challenge-of-paying-for-smart-cities-projects.pdf

El Emam, K., Rodgers, S., & Malin, B. (2015). Anonymising and sharing individual patient data. *BMJ (Clinical research ed.), 350.*

Fishman, T. D., & Flynn, M. (2018). Part two: Funding and financing smart cities series. Using public-private partnerships to advance smart cities. Deloitte, 1–10. Retrieved from www2.deloitte.com/content/dam/Deloitte/global/Documents/Public-Sector/gx-ps-public-private-partnerships-smart-cities-funding-finance.pdf

Fleming, S. (2019, 19 February). Commuters in these cities spend more than 8 days a year stuck in traffic. World Economic Forum. Retrieved 20 January

2021 from www.weforum.org/agenda/2019/02/commuters-in-these-cities -spend-more-than-8-days-a-year-stuck-in-traffic/

Forzieri, G., Bianchi, A., Silva, F. B. e., Marin Herrera, M. A., Leblois, A., Lavalle, C., Aerts, J. C. J. H., & Feyen, L. (2018). Escalating impacts of climate extremes on critical infrastructures in Europe. *Global Environmental Change, 48*, 97–107.

Gambella, C., Monteil, J., Dekusar, A., Cabrero Barros, S., Simonetto, A., & Lassoued, Y. (2019). A city-scale IoT-enabled ridesharing platform. *Transportation Letters*, 1–7.

Gascó-Hernandez, M. (2018). Building a smart city: Lessons from Barcelona. *Communications of the ACM, 61*(4), 50–57.

Gehl, J. (2013). *Cities for People*. Island Press.

GovTech Singapore. (2017, 21 September). On the road with IoT. GovTech. Retrieved 20 January 2022 from www.tech.gov.sg/media/technews/on-the -road-with-iot

Gutiérrez Escolar, A., Castillo Martínez, A., Gómez, J., Gutiérrez-Martinez, J.-M., Stapic, Z., & Medina, J. (2015). A study to improve the quality of street lighting in Spain. *Energies, 8*, 976–994.

Huang, T.-J. (2017). Imitating the brain with neurocomputer: A 'new' way towards artificial general intelligence. *International Journal of Automation and Computing, 14*(5), 520–531.

IDC. (2018, 23 July). IDC forecasts smart cities spending to reach $158 billion in 2022, with Singapore, Tokyo, and New York City among top spenders. Business Wire. Retrieved 4 January 2021 from www.businesswire.com/news/ home/20180723005083/en/IDC-Forecasts-Smart-Cities-Spending-to-Reach -158-Billion-in-2022-with-Singapore-Tokyo-and-New-York-City-Among -Top-Spenders

IEA. (2017). World energy outlook 2017: China. International Energy Agency. Retrieved 1 February 2022 from www.iea.org/weo/china/

Inhabitat. (2011, 27 May). Fujisawa Smart Town planned for Japan to be most advanced eco city in the world. Inhabitat. Retrieved 21 January 2021 from https://inhabitat.com/fujisawa-smart-town-planned-for-japan-to-be-most -advanced-eco-city-in-the-world/

Ismagilova, E., Hughes, L., Rana, N. P., & Dwivedi, Y. K. (2020). Security, privacy and risks within smart cities: Literature review and development of a smart city interaction framework. *Information Systems Frontiers*. https://doi .org/10.1007/s10796-020-10044-1

Jacobs, J. (1961). *The Death and Life of Great American Cities*. Random House.

Kamel Boulos, M. N., Tsouros, A. D., & Holopainen, A. (2015). 'Social, innovative and smart cities are happy and resilient': Insights from the WHO EURO 2014 International Healthy Cities Conference. *International Journal of Health Geographics, 14*(1), 3.

Kolotouchkina, O., & Seisdedos, G. (2017). Place branding strategies in the context of new smart cities: Songdo IBD, Masdar and Skolkovo. *Place Branding and Public Diplomacy, 14*(2), 115–124.

Kulp, S. A., & Strauss, B. H. (2019). New elevation data triple estimates of global vulnerability to sea-level rise and coastal flooding. *Nature Communications, 10*(1), 4844.

KUNA. (2018, 15 March). KISR: Shagaya project first step toward energy diversification. Kuwait News Agency. Retrieved 2 February 2019 from www.kuna.net.kw/ArticleDetails.aspx?id=2703316&language=en

Liu, S. (2019, 24 October). Global IoT market size 2017–2025. Statista. Retrieved 13 December 2019 from www.statista.com/statistics/976313/global-iot-market-size/

Maisels, C. K. (1993). *The Emergence of Civilization: From Hunting and Gathering to Agriculture, Cities, and the State in the Near East.* Routledge.

Malanga, S. (2018, Summer). The promise and peril of 'smart' cities. *City Journal.* Retrieved 23 January 2022 from www.city-journal.org/promise-and-peril-smart-cities-16035.html

Martinez-Balleste, A., Perez-Martinez, P., & Solanas, A. (2013). The pursuit of citizens' privacy: A privacy-aware smart city is possible. *IEEE Communication Management, 51*(6), 136–141.

McCarthy, N. (2020, 10 March). *Traffic Congestion Costs US Cities Billions of Dollars Every Year.* Forbes.

McKinsey & Company. (2018). Smart city solutions: What drives citizen adoption around the globe? Retrieved 1 February 2022 from www.mckinsey.com/industries/public-sector/our-insights/smart-city-solutions-what-drives-citizen-adoption-around-the-globe

Mohamed, N., Al-Jaroodi, J., Jawhar, I., Idries, A., & Mohammed, F. (2020). Unmanned aerial vehicles applications in future smart cities. *Technological Forecasting and Social Change, 153*, 119293.

Mordor Intelligence. (2021). Smart cities market: Growth, trends, COVID-19 impact, and forecasts (2022–2027). Retrieved 20 January 2022 from www.mordorintelligence.com/industry-reports/smart-cities-market

Nastu, P. (2020). IT could cut global emissions 15%, saving $800 billion. *Environment Leader.* Retrieved 21 January 2021 from www.environmentalleader.com/2008/06/smart-2020-it-could-cut-global-emissions-15-saving-800-billion/

Newman, P., & Kenworthy, J. (2015). *The End of Automobile Dependence: How Cities Are Moving beyond Car-Based Planning.* Island Press.

Nižetić, S., Šolić, P., López-de-Ipiña González-de-Artaza, D., & Patrono, L. (2020). Internet of Things (IoT): Opportunities, issues and challenges towards a smart and sustainable future. *Journal of Cleaner Production, 274*, 122877–122877.

Oughton, E. J., Frias, Z., van der Gaast, S., & van der Berg, R. (2019). Assessing the capacity, coverage and cost of 5G infrastructure strategies: Analysis of the Netherlands. *Telematics Informatics, 37*, 50–69.

Perwej, Y., & Parwej, F. (2012). A neuroplasticity (brain plasticity) approach for use in artificial neural network. *International Journal of Scientific and Engineering Research, 3*(6), 1–9.

Peters, M. A., & Besley, T. (2019). 5G transformational advanced wireless futures. *Educational Philosophy and Theory*, 1–5. https://doi.org/10.1080/00131857.2019.1684802

Porphyrios, D., Robertson, J., & Rowe, C. (1985). *Leon Krier: Houses, Palaces, Cities: An Architectural Design Profile*. St Martin's Press.

Prensa, N. d. (2019). Artificial intelligence and VAR take centre stage during LaLiga Innovation Showcase at the Mobile World Congress. LaLiga. Retrieved 23 January 2022 from www.laliga.com/en-GB/news/artificial-intelligence-and-var-take-centre-stage-during-laliga-innovation-showcase-at-the-mobile-world-congress

Rathee, G., Khelifi, A., & Iqbal, R. (2021). Artificial intelligence (AI-)enabled Internet of Things (IoT) for secure big data processing in multihoming networks. *Wireless Communications and Mobile Computing*, *2021*, 5754322.

Ristvej, J., Lacinák, M., & Ondrejka, R. (2020). On smart city and safe city concepts. *Mobile Networks and Applications*, *25*(3), 836–845.

Rodriguez-Segura, D. (2020). *Educational Technology in Developing Countries: A Systematic Review*. EdPolicy Works working paper. University of Virginia. Retrieved from www.curry.virginia.edu/sites/default/files/uploads/epw/72_Edtech_in_Developing_Countries.pdf

Rohaidi, N. (2018, 16 November). Singapore wins Smart City of 2018 award. GI. Retrieved 5 March 2019 from https://govinsider.asia/smart-gov/singapore-wins-smart-city-of-2018-award/

Rumsey, A. B. (2022). The infrastructure investment and jobs act – half a billion dollars in 'smart city' investments. Arnold & Porter. Retrieved from www.arnoldporter.com/en/perspectives/blogs/environmental-edge/2022/01/half-a-billion-dollars-in-smart-city-investments

Safiullin, A., Krasnyuk, L., & Kapelyuk, Z. (2019). Integration of Industry 4.0 technologies for 'smart cities' development. *IOP Conference Series: Materials Science and Engineering*, *497*, 012089.

Salingaros, N. A. (2000). Complexity and Urban Coherence. *Journal of Urban Design*, *5*, 291–316.

Samuel, R. D., Galily, Y., Filho, E., & Tenenbaum, G. (2020). Implementation of the video assistant referee (VAR) as a career change-event: The Israeli Premier League case study. *Front Psychol*, *11*, 564855.

Selzer, S., & Lanzendorf, M. (2019). On the road to sustainable urban and transport development in the automobile society? Traced narratives of car-reduced neighborhoods. *Sustainability*, *11*(16). https://doi.org/10.3390/su11164375

Sharma, M., Joshi, S., Kannan, D., Govindan, K., Singh, R., & Purohit, H. C. (2020). Internet of Things (IoT) adoption barriers of smart cities' waste management: An Indian context. *Journal of Cleaner Production*, *270*, 122047.

Smart Cities Association. (2021). *Global smart cities market to reach a whopping $3.5 trillion by 2026*. Smart Cities Association. Retrieved 20 January 2021 from www.smartcitiesassociation.org/index.php/media-corner/news/1-global-smart-cities-market-to-reach-a-whopping-3-5-trillion-by-2026

Soja, E., Morales, R., & Wolff, G. (1983). Urban restructuring: An analysis of social and spatial change in Los Angeles. *Economic Geography*, *59*(2), 195–230.

Spitz, J., Wagemans, J., Memmert, D., Williams, A. M., & Helsen, W. F. (2021). Video assistant referees (VAR): The impact of technology on decision making in association football referees. *Journal of Sports Science, 39*(2), 147–153.

UN Habitat. (2020). Global state of metropolis 2020. Retrieved 1 February 2022 from https://unhabitat.org/sites/default/files/2020/06/gsm-population-data-booklet2020.pdf

UNCTAD. (2020). The least developed countries report 2020. UN Publications. Retrieved from www.worldbank.org/en/topic/tertiaryeducation

UNCTAD. (2021). Technology and innovation report 2021: Catching technological waves innovation with equity. UNCTAD. Retrieved from https://unctad.org/system/files/official-document/tir2020_en.pdf

UNEP. (2014). Emerging issues for small island developing states. UNEP. Retrieved from https://sustainabledevelopment.un.org/content/documents/1693UNEP.pdf

UNFCCC. (n.d.). Vulnerability and adaptation to climate change in small island developing states. UNFCCC. Retrieved 8 August 2021 from https://unfccc.int/files/adaptation/adverse_effects_and_response_measures_art_48/application/pdf/200702_sids_adaptation_bg.pdf

United Nations. (2020). Cities and pollution. UN. Retrieved 1 November 2021 from www.un.org/en/climatechange/climate-solutions/cities-pollution

United Nations Department of Economic and Social Affairs. (2019, 27 September). Small island developing states, On the front lines of climate and economic shocks, need greater international assistance. Retrieved 13 August 2021 from www.un.org/development/desa/en/news/sustainable/sids-on-climatechange-front-line-need-more-assistance.html

Vailshery, L. S. (2021, 8 March). IoT an non-IoT connections worldwide 2010–2025. Statista. Retrieved 21 January 2022 from www.statista.com/statistics/1101442/iot-number-of-connected-devices-worldwide/

Woetzel, J., Remes, J., Boland, B., Katrina, L., Sinha, S., Strube, G., Means, J., Law, J., Cadena, A., & Tann, v. d. V. (2018). Smart cities: Digital solutions for a more liveable future. McKinsey Global Institute. Retrieved 1 February 2022 from www.mckinsey.com/business-functions/operations/our-insights/smart-cities-digital-solutions-for-a-more-livable-future

Yao, X., Farha, F., Li, R., Psychoula, I., Chen, L., & Ning, H. (2021). Security and privacy issues of physical objects in the IoT: Challenges and opportunities. *Digital Communications and Networks, 7*(3), 373–384.

Yu, Z., Zhang, H., Tao, Z., & Liang, J. (2019). Amenities, economic opportunities and patterns of migration at the city level in China. *Asian and Pacific Migration Journal, 28*(1), 3–27.

Zoonen, L. v. (2016). Privacy concerns in smart cities. *Government Information Quarterly, 33*(3), 472–480.

2. The underlying and basic foundations of the smart city: where do artificial intelligence, machine learning and other buzz words fit in the narrative?

INTRODUCTION

The smart city concept is made up of an intricate and complex integration of different smart technologies, where its heart and soul comprise of the numerous data being generated, processed and analysed at different levels. Like an animal's body that is made up of different organs each playing a distinctive function, the smart city concept comes about as a result of a combination of different technologies, each playing a significant part with respect to data generation, storage, transfer, processing, computation, analysis and decision making (Allam, 2021; Allam & Dhunny, 2019). Therefore, technologies being deployed in smart cities could be categorised according to the role they play in different stages of the concept implementation, more so in regard to data management and the security of both the physical hardware and software. The first and most crucial category comprises technologies that have the capacity to help in data generation, which has been argued to be the beginning of big data (Kong et al., 2020). Technologies involved in this category include the Internet of Things (IoT), artificial intelligence (AI) and mobile communication technologies like 4G, 5G and the much anticipated 6G (Allam & Jones, 2021; Elmeadawy & Shubair, 2019; Nguyen et al., 2021).

IoT technology, in this case, comprises the network of physical objects (things) that are integrated (embedded) in different devices such as sensors, cameras, wearables and others. Also included in IoT are software and networks that are connected in such a way that they have capacities to exchange data in real time in a given central network (Adly et al., 2020; Gill et al., 2019; Nižetić et al., 2020). AI technology on the other hand is associated with rendering the IoT devices and networks 'intelligent'; hence allowing for real-time data generation, data transfer, analysis and the drawing of different insights. It also allows for connectivity of different devices that are within a network and, with the help of mobile communication technologies, make it possible for faster information transfer (Peters & Besley, 2019). IoT and AI technologies further play a critical role in facilitating the generation of sufficient and quality data that would warrant far-reaching and concrete decisions on different issues pertaining to cities (Bibri, 2021b; Cech et al., 2018).

The second category that is vital to the smart city concept comprises of all technologies that allow for data storage, data computation and analysis and decision making following insights drawn from the analysis. Technologies such as big data, cloud computing, machine learning, AI and blockchain are critical for these purposes (Allam, 2019a; Allam & Dhunny, 2019). From a wide range of studies (IDC, 2020; Research and Markets, 2021), one main challenge that smart cities are facing is data storage following an exponential increase of the amount of data being generated. For instance, it is anticipated that by 2025, the amount of data generated will be more than 180 zettabytes, an over 100 per cent increase from the 79 zettabytes recorded in 2021 (See, 2021). This growth is in part credited to the exponential growth of IoT devices that are also experiencing an explosion in demand. The analysis and computation of these data, however, is argued to be most critical as it allows cities to experience changes like automation that are critical in improving efficiency and performance. With the aforementioned technologies, it is not surprising that some cities are now gradually becoming hosts to autonomous vehicles, increasing e-commerce activities, have managed to install safe cycling lanes and are able to effectively monitor and control traffic (Biyik et al., 2021; Dixit et al., 2021). The strength of smart technologies was in particular experienced during the COVID-19 lockdowns, where services like delivery from stores in some cities spread across different geographical locations were carried out using drones. A case in point is Zapallar, Chile, where drones were deployed to deliver medicine to the elderly (Gulistan News, 2020). During the same period, cycling became essential for many urban residents (De Vos, 2020; Deutsche, 2021; Vandy, 2020), providing them with much needed health benefits, cementing the transport mode as an alternative mobility contender and health tool that can help reduce car dependency, reduce traffic and promote healthy lifestyles in cities.

Technologies such as blockchain are becoming popular in smart cities, not only in helping actualise transactions, especially through crypto-currencies (Allam & Jones, 2019c), but even more in the governance of cities (Allam, 2018; Shahab & Allam, 2019). After big data have been analysed and decisions made, blockchain technologies can help in actualising them by allowing for safer interconnections, especially of most urban vertical services such as security, energy, mobility and health (Iberdrola, 2021). In part, this is achieved as blockchain technology helps in securing databases by making them immutable. This characteristic can then be tapped to allow for open access to data and information in real time for all stakeholders within the city network. It also increases the trust components in cities, especially for contracting parties within urban milieus, thus making it possible for technologies such as peer-to-peer to be deployed, even within small-scale activities like trading excess solar energy produced from rooftops (IRENA, 2020; Zhang et al., 2018). Within the confines of the smart city concept, it has been established that one critical concern especially with residents is on how data and private information are utilised (Allam, 2019b, 2020a). These concerns have escalated such that it becomes challenging for a sizeable number of resi-dents to share or participate in activities that would require them to avail their data. However, blockchain technology has the potential to alleviate these fears, allowing for a high degree of reliability, transparency and immutability (Ismagilova et al., 2020; Shrestha et al., 2020). Further, it eliminates the need for third parties, which is critical for the privacy and security of data.

Overall, it is evident that smart cities cannot exist without the numerous and diverse technologies being deployed, and the prospect of even more technologies arising in the future is critical for the 'smartness' of the city. Already, with existing technologies, it is evident that the future of most cities, especially with the challenges of climate change, urbanisation and pandemics like COVID-19, relies on the extent of the deployment and utilisation of those technologies (Bibri, 2021a). Without these, as will be discussed in the succeeding sections, most cities have few buffers against modern-day urban challenges, and most will succumb to outcomes such as climate migration, heatwaves, flooding, drought and other climate events that are on the rise. Cities will further continue to experience unprecedented economic losses as a result of traffic congestion, reduced resources, unsustainable practices and other issues that have been found to be comprehensively addressed by the deployment of technologies in cities (Afrin & Yodo, 2020). In light of this, this chapter seeks to explore the role and impacts of different technologies as key in the unfolding of the smart cities concept.

CITIES WITHOUT TECHNOLOGIES

Cities across the globe have been identified as the main engine of economic growth, especially with their economic contribution approximated to reach 70 per cent of the total global gross domestic product (GDP) (Dobbs et al., 2015). The contribution of cities has been consistent, even before the Industrial Revolution that marked the official period when contemporary cities began to emerge. Ancient cities (from around 1100 CE to 6500 BCE; World History Encyclopedia, 2021) were seen as the melting pots for culture and socialisation, as centres for higher learning, entertainment and recreation (Chametzky, 1989). However, while those roles continued to increase and cities' influence in society continued to gain roots, they started to experience varied challenges. Cities began to experience increased cases of insecurity and notable social and economic inequalities – for example, insufficient housing, scarcity of basic resources like food and water and others have been common in cities (Chametzky, 1989). Since the emergence of the third industrial revolution, more challenges like rapid population growth, climate change incidences, traffic congestion and land scarcity have become common. Other challenges include waste management, insufficient or poor administrative services delivery and pandemics.

Before the emergence and subsequent adoption of different technologies as options to solve the diverse urban challenges, the solutions adopted could be argued to have been insufficient. That is, while they were being fronted as answers for the different urban problems, they opened doors for more challenges, hence leading to cyclical problem solving. For instance, as a solution to the increasing challenge of the lack of housing and increasing costs of land, it is reported that some cities across the globe experienced unprecedented levels of urban sprawl (Camagni et al., 2002; Duany et al., 2000). In part, this helped to reduce the number of urban residents who were caught up in the housing shortages. It also allowed people, especially in the emerging middle-class cluster, to own their own homes rather than reside in the communities that were popular in the suburban areas. However, those trends opened doors for challenges like insufficient service delivery especially in the distribution of basic services such as sewerage, water and energy distribution and waste management (Moe & Wilkie, 1999). The urban sprawl is further associated with an increased demand for basic infrastructure such as a transport system, education and health facilities, public markets and others. Those have somehow contributed to some local governments plunging into debt as they endeavour to increase the provision of different basic infrastructures and services. On the social front, while most compact cities are criticised for inequality and gentrification (where minorities and the poor are forced out of their neighbourhoods due to factors like the higher cost of living and high cost of land; Atkinson, 2000), the urban sprawl is accused of intensifying social segregation as it allows people to live solitary lifestyles (in their own compounds) (Hortas-Rico & Solé-Ollé, 2010). It further perpetuates and contributes to traffic congestion as people are forced to own automobiles to access different basic urban nodes that are sparsely located in sprawled urban areas.

In cities where the sprawl has been contained, or land is limited, urban regeneration has been adopted as a viable solution. On this, where different technologies are being adopted to assist in regeneration, the outcomes have substantial benefits. However, that is not the case in areas where regeneration is undertaken in the absence of technologies (Atkinson, 2000). In the latter scenario, most programmes are uncoordinated and expensive as they are not informed or backed by any credible data-driven insights. The end results of most such programmes are trails of numerous ghost mega projects: abandoned programmes or cities that have very low liveability index scores. Another critical outcome of urban regeneration undertaken without the deployment of technology could be the negation of cultural heritage sites and culture itself, especially where gentrification happens, thus destabilising existing social and community order and networks (De Cesari & Dimova, 2019).

The sustainable concerns in most cities are by a higher margin associated with energy production and consumption, with the overreliance on fossil fuels being a major concern (Allam, 2019c). Due to household consumption, industries and in transportation sectors, and with the urban population continuing to grow at an unprecedented rate, more energy is demanded. Demand is also prompted by an increased growth in the number of automobiles, which are reported to have reached over 1.4 billion by the end of 2019 (Okeafor, 2020). The cumulative impacts of demand for energy in different frontiers is an increase in emissions from energy, which is reported to have accounted for 73.2 per cent of all emissions globally as of 2020 (Ritchie & Roser, 2021). On this, it is paramount to note that due to infiltration of some technologies in the energy sector, working in favour of increased sustainability, some alternative energy sources such as solar, wind and hydro have been supplementing fossil fuels. Therefore, where these alternatives were absent, the 73.2 per cent reported will have been relatively higher, thus affirming the importance of different technologies in cities.

Technology deployment in urban conservation has been observed to offer opportunities to lower operational costs, hence prompting cities to invest in digital solutions, even if they represent higher upfront costs. Conservation, especially of buffer zones (forests, wetlands, coastal vegetation like mangroves, etc.), is associated with numerous benefits such as improvement in air and water quality, beautification of landscapes, protection of soil resources and enhancement of wildlife and aquatic animals (Diener & Mudu, 2021). Buffer zones further help in the regulation of urban temperatures and thus mitigate extreme conditions (Aram et al., 2019). In cities where such conservation is not hinged on the use of technologies, the chance of not achieving desirable outcomes is high. These are then followed by incidences such as erratic weather conditions, increased disease outbreaks and increased accidents associated with infrastructure collapse (United Nations Framework Convention on Climate Change, 2015) due to poor soil prompted by erosion, flooding and geotectonic activities. With technology use for conservation, even new and advanced approaches like the adoption of green roofs and walls could be used to complement already existing green spaces. The adoption of such new approaches has been associated with a reduction of heat islands to a margin of 5 degrees Celsius (United States Environmental Protection Agency, 2020). Besides temperature, technology can enhance air quality, and this is critical for cities as most of them are already facing the challenge of pollution from transport and the manufacturing waste sector (United Nations, 2020).

It is worth noting that technology is at the forefront of the achievement of the Sustainable Development Goals, especially Goal 11, the New Urban Agenda (United Nations, 2017) and other global-scale policy frameworks (United Nations, 2011) that call for improvement of situations in urban areas. Therefore, in the quest of building cities that incorporate and support human-scale attributes, technology deployment is inevitable, in particular, smart technologies that have the capacity to transform performance and increase efficiency of different aspects of life in cities.

EMERGING TECHNOLOGIES AND THEIR POTENTIAL IN THE SMART CITY CONCEPT

The role of cities in the global, regional, national and local economies has been widely documented, with an almost universal agreement about how much they contribute to the global GDP. It is worth noting that cities across the globe differ greatly in terms of size, GDP, population and infrastructural investments. However, despite their attributes, they are key to national socioeconomic endeavours. This could be attributed to the fact that some cities are far larger in terms of size and economic growth than many countries. For instance, Tokyo-Yokohama in Japan has a total population of 37.7 million people (Population U, 2022). If computed against national population size, it would rank in position 40 globally (World Meter, 2022). Tokyo is also the richest city in terms of GDP (at $1.602 billion as of 2020) while New York ranks second with a GDP of $1.561 billion. Tokyo could therefore be in position 13 if it were to be ranked in the list of countries by GDP size (Population U, 2022). It would be followed in position 14 by New York, which is above Spain ($1.439.96 billion as of 2021) in terms of GDP (International Monetary Fund, 2021). These examples show that cities are at the centre of global economic and social agendas. But, the most critical underlying factor that has made most of the top-ranked cities in terms of GDP experience such growth is the commitment in the deployment of diverse technologies to offer solutions for urban challenges. While technologies alone have no capacity to alleviate the challenges, their incorporation in cities to complement human efforts could be argued to allow most cities to maintain a competitive edge as well as improve performance.

In these cities, the deployment of technologies such as AI, IoT and big data has been given priority and sufficient financial resources. For instance, in Tokyo, investment in AI technologies (especially in the subdomain of robotics) has been gaining momentum particularly with an aim of complementing the human contribution with machine labour. This option is embraced as a caution to the notable trend of shrinking in the labour force prompted by an ageing population. Projections are that over half of the job opportunities will be filled by robots and machines by 2035 in Tokyo-Yokohama (Gonzalez, 2021). In New York, smart technologies are reported to have become part of the urban fabric, accounting for over 291,000 jobs already created and yielding approximately $124.7 billion to the city's GDP (NYC, 2020). In Singapore City, smart technologies such as AI, machine learning and big data have been deployed in different sectors (Allam, 2020d, 2020e), including water, where the country is now able to provide over 40 per cent of its water from recycling plants (Taylor, 2019) rather than importing from neighbouring countries, as has been the norm. In other cities like Amsterdam, the presence of smart technologies has allowed the city to cut a niche for itself in terms of branding and marketing. As a result, the city is yielding almost five-fold returns on every unit of technological investment through increased tourism activities taking place in the city (Iamsterdam, 2021).

The different smart technologies now available for deployment in cities are reported to have given cities alternatives in terms of positioning and competitive advantage. As with the case of Amsterdam noted above, other cities such as Singapore, Kuala Lumpa, Barcelona, London and Paris to name just a few are using the availability of technologies to position themselves as tourism hotspots (Gascó-Hernandez, 2018; Zvolska et al., 2019). Using smart technologies, cities have also been reported to be striving to rank high in terms of liveability status (Allam, 2020b; Dabeedooal et al., 2019). For instance, in Singapore, there is an intensified use of smart traffic cameras that help in monitoring and controlling different aspects of traffic, including helping reduce congestion, allowing for the smooth flow of traffic (Allam & Jones, 2018). In Paris, smart technologies are being utilised to help the city reduce the number of private cars in the city and, as an alternative, help create an environment conducive for cyclists (Coldwell, 2014; Mancebo, 2020). The outcome of these initiatives have seen the city reduce its traffic while at the same time increasing the number of bicycles, especially by providing opportunities to startups to help in building digital solutions (for instance, Velib, a ride-sharing service provider, introduced over 14,000 bicycles in the city in 2014; Albert, 2019).

The advancement in smart technologies is further providing cities with the capacity to invest in and emphasise alternative energies such as solar and wind. Cities are now able to deploy products such as smart meters, smart housing projects and smart grids to optimise their energy usage. These products in turn have had significant impacts on increasing energy efficiency, thus helping reduce demand as well as overconsumption of non-renewable resources like fossil fuels, which have long been the primary source of energy production. Emerging technologies such as blockchain and peer-to-peer have further opened opportunities for urban residents to create alternative income streams, via options like the sale of developmental air rights (Allam & Jones, 2019c) or surplus energy from small-scale solar energy production (Alam et al., 2017). The surplus energy is either bought by neighbours or by local governments. Such opportunities not only help to create extra income for locals, but they also allow cities to achieve their objectives of reducing carbon footprints. Additionally, such initiatives are worthwhile for the role that emerging technologies are playing in making the basic components of solar panels and batteries relatively cheaper, hence allowing alternative energies to compete with conventional energy production methods that rely on non-renewable sources (IRENA, 2017, 2019).

Other technologies such as 3D printing, digital twins, extended reality and robotics are also slowly gaining traction in city planning, especially in areas like automation, security enhancement and housing programmes (through pre-fabrications). These are helping stakeholders in the urban realm to address and manage emergency cases much faster, more efficiently and at a reduced cost (Rathore et al., 2021). While smart city projects are commonly perceived as being expensive undertakings, requiring huge initial capital investment, they can provide cities with the capacity to increase revenue through cheaper regeneration programmes and a wide range of new economic opportunities. With this, it is arguably important for collaboration between local governments, national government and other agencies to finance smart projects in cities (Allam & Jones, 2019b; Allam et al., 2020).

THE HARMONY OF DIFFERENT TECHNOLOGIES IN THE SMART CITY AGENDA

The aspiration to transform existing and emerging cities to become 'smart' is a novel step toward addressing a majority of urban challenges. However, it is worth noting that technology cannot be viewed as a silver bullet for all major urban challenges; timely human interventions are also necessary. While there are successful cases, a myriad of practical examples have highlighted that there still remain some issues that require ironing out. This is critical before it can be positively argued that technology is the ultimate single tool or solution to address the urban fragments caused by contemporary planning, or societal challenges. A case in point is the city of Santander in Spain, which in 2009 installed over 12,000 sensors on its different establishments (Newcombe, 2014). Unfortunately, these were installed even before the prerequisite technologies such as IoT and AI had fully matured to a point of being deployed in urban areas; hence resulting in an untimely failure of the anticipated smart city plan.

Another reference is the Smart City Songdo in South Korea. Songdo can be termed as a practical example of what many cities across the world should emulate. The city has perfectly networked computers that operate as central servers for sensors, cameras and other smart devices installed therein to help monitor diverse aspects such as climate, traffic and communications (Eireiner, 2021). These work well, but the city has fallen short in attracting residents and businesses. From reliable reports, the city has managed to attract only one-third of its 300,000 capacity, and critics argue that despite the elaborate 'smartness', the city's social dimensions are far below par and this needs to be addressed (Albert, 2019).

The two examples above affirm calls by Salingaros (2000) for cities to be planned in such ways that that guarantee complex order and coherence. In the above cases, it is evident that attention by different stakeholders was more on the technology, sustainability and economic dimensions, while the social dimension was overlooked or underemphasised. Therefore, in the pursuit of smart agendas, different technologies need to be deployed in a manner that allows for the achievement of perfect harmony that could ultimately help cities become 'whole', as envisioned by Alexander et al. (1977). The starting point on this could be in ensuring that all technologies, networks and smart devices are seamlessly integrated to communicate with each other. That is ensuring that there is standardisation of protocols instead of each smart city service provider maintaining their own unique network. Without such standardisation, it is possible that cities could maintain an unprecedented number of contracted service providers, thus making it difficult to ensure that all dimensions that make up the city's fabric are being addressed adequately (Allam & Jones, 2020, 2021).

One major concern that arises from the disharmony in the integration of smart technologies in cities mostly perpetuated by large corporations is that of data usage, privacy and security (Ismagilova et al., 2020). In most cases, a majority of local governments are reported to contract third-party service providers to manage data, and this has been a source of concern. This is particularly attributed to the profit-centric nature of many service providers, where some have been accused of monetising data or using data for purposes beyond the contract limits (Allam, 2019b, 2020c). One example is the case of Sidewalk Toronto (McGrath, 2020), a smart city project that failed to materialise due to data and privacy concerns raised by locals. Sidewalk Labs was contracted to offer smart services, but most of the local citizens were not happy, especially with how their data would be managed or applied, and thus they objected to the programme (Fussell,

2018). This example amplifies the need for local government to create their own capacity for managing diverse digital urban aspects, especially with data, so as to allow them to successfully operate smart city projects.

The harmonisation of different technologies in smart cities can further be achieved by ensuring there is sufficient engagement between diverse stakeholders. In particular, urban residents have often been neglected in decision making, including on projects that directly impact them (De Guimarães et al., 2020; Pereira et al., 2017). This creates an environment of mistrust and apprehension, where residents openly oppose certain projects that they are not acquainted with from the beginning. A case in point is the modern suburbia project that was undertaken in the United States involving the development of Levittown, Reston and Celebration (Marshall, 2015). Although the project was not directly funded through public resources but involved a partnership between local governments and the private sector, it was marred by accusations of a lack of public participation. The project was seen as a plot aimed at excluding, segregating and persecuting residents (Galyean, 2015). However, proponents of the project argued that the intent was noble and would have allowed the adoption of efficient methods like fabrications which would ultimately have reduced the cost of houses. In a different scenario that showed the importance of engaging different stakeholders, the city of Bogota, Colombia successfully managed to redesign its urban morphology by incorporating green aspects like bicycle lanes (Chatterton & Hanway, 2021). The redesign first underwent robust public participation where residents were engaged via an open-source online platform. By the time the project was undertaken residents were already on board, thus leading to its success in reducing traffic bottlenecks and increasing mobility (Chatterton & Hanway, 2021).

As a way of affirming the importance of wide participation in urban transformation projects, UN-Habitat launched a flagship programme in 2020 dubbed 'People Centered Smart Cities' (UN-Habitat, 2020). This programme acknowledged that it is inevitable that cities across the globe will adopt digital solutions to transform diverse urban aspects. However, these need not exclude stakeholders, especially residents who are key beneficiaries of projects such as smart housing, smart traffic and smart grids. Collaboration between stakeholders such as planners, designers, financiers and others would help overcome challenges like non-standardised protocols. Public participation and robust collaboration should be a must in most urban public projects.

URBAN CHALLENGES ADDRESSED BY TECHNOLOGIES

A number of urban challenges including climate change, traffic congestion, increasing urban population and socioeconomic inequalities are not unique to certain cities, but are a bane to most if not all cities. This explains why there have been calls, including during global summits as captured in documents such as Sustainable Development Goal 11 (UNEP, 2015), *New Urban Agenda* (United Nations, 2017) and the 2030 Agenda for Sustainable Development (United Nations, 2011) for urgent action. Climate change, for instance, has been identified as a major twenty-first-century challenge that unless amicably addressed would prompt dire consequences on many economies. To put this into perspective, in the recent past, numerous challenges associated with direct impacts of climate change have been observed across cities and economies. For instance, in the small island developing states, climate change has prompted a myriad of challenges that have threatened not only the livelihoods of the locals, but also their habitats, infrastructures, lives of residents and in worst-case scenarios total submersion of some states (Allam & Jones, 2019a). An example is Kiribati and Tavula that are experiencing unprecedented rise in sea levels (UNEP, 2014). In developed countries, different cities are experiencing increased cases of flooding, heatwaves, erratic precipitation and changing seasons (longer summers and shorter winters). While there are numerous proposals that have been formulated and adopted, it has become apparent that it might not be tenable to achieve projected sustainable agendas. For instance, there are doubts on the ability of the world to ensure that global temperatures do not rise beyond the 2 degrees Celsius anticipated in the Paris Agreement (United Nations Framework Convention on Climate Change, 2015). Failing to meet this object will have far-reaching impacts on most global cities that are already strained in terms of resources by climate change incidences. As a way to help find solutions that might complement approaches already being adopted in cities, proponents of the smart city concept believe that integration of smart technologies could act as a positive agent. For instance, it is argued that technologies such as AI, IoT and machine learning could help in areas like prediction of future climate scenarios, hence prompting better-informed measures and decisions.

In the cause of deploying technologies to address endemic urban challenges like issues in the traffic sector, it is argued that the benefits would spill over including to climate change mitigation programmes. For example, adopting technologies that allow for an increased deployment of electric vehicles result in a reduction in consumption of non-renewable energy products in different transport fields and help decrease the sector's carbon footprint (Biyik et al., 2021; Dixit et al., 2021). Critics of electric vehicles argue that their introduction is a 'zero-sum game' as they would continue to consume electricity generated from non-renewable sources (Fogelberg, 1998). However, it has been established that most electric vehicles being produced and those already on the roads rely heavily on renewable energy or natural gas. Therefore, they contribute greatly to the sustainable agenda, and that it is a key stepping stone in the transition to a fully renewable future. The use of renewable energy in vehicles is also evident with hybrid electric vehicles that can be designed to only rely on fuel in extreme situations (Himelic & Kreith, 2008; Weiss et al., 2019). Besides enhancing energy efficiency and reduced consumption, smart technologies in urban areas also enhance the transport sector by helping reduce traffic congestion. This is saving cities from increased economic losses from both lost working hours and extra resources consumed when cars are stuck in traffic. On this, it is reported that in the state of Los Angeles alone, over $8 billion was lost due to traffic congestion, while in Chicago a total of $7.6 billion was also lost in 2019 alone. Cumulatively, over $66.1 billion was lost in different urban areas in the United States in 2019 (McCarthy, 2020).

One of the most visible transformations that cities across the globe are reported to be undergoing is urban regeneration, targeting a number of aspects such as conservation of heritage areas (Appendino, 2017; Riganti, 2017), expansion of infrastructure (Kim et al., 2020) and restructuring of housing programmes (Mercader-Moyano & Serrano-Jiménez, 2021). The attention on regeneration is prompted by the twin phenomenon of rapid urbanisation and the increasing urban population (Allam & Newman, 2018). Regeneration is further being undertaken in cities as a way of helping address inequality through redistribution of opportunities as well as helping to increase prosperity and quality of life in areas that might have previously been neglected (Kim et al., 2020). Such programmes have facilitated many cities to improve their status; moving higher in rank in aspects like liveability, security and attractiveness. A case in point is Singapore, in which elaborate urban transformation and regeneration programmes allowed the island state to change its status from a developing country to among the most liveable cities in the world (Allam, 2020f; Crinson, 2017).

Urban regeneration programmes and initiatives are complex, expensive and lengthy and sometimes attract negative outcomes such as gentrification. However, deployment of technology is helping overcome some of the challenges associated with regeneration. For instance, in the case of culture and cultural heritage, modern technologies are helping in the digitisation of cultural objectives, entities and items, thus preventing them from being lost (Adzaho, 2013). Technologies such as digital twin and virtual reality can help store cultural objects in digital form, hence making sure that even in cases of destruction of the physical forms they can be duplicated using technologies such as 3D printing (Balletti & Ballarin, 2019). Use of modern technology also facilitates museums and cultural centres to incorporate virtual products, thus giving access to a wide scope of visitors. This is critical in the quest to increase cities' income streams as well as ensuring perpetuity of different artefacts that could otherwise be lost over time (Nikonova & Biryukova, 2017). In projects like the regeneration of old houses, deployment of technology can help achieve green status in terms of smart energy, water conservation, green roofing and green walls.

Going into the future, while it will take time to alleviate all major urban challenges, the deployment of smart technologies is expected to be part of diverse urban solutions. Already much progress and many benefits have been achieved, and with proper and concerted efforts, coupled with sufficient funding, complementing urban management and governance with technologies will yield numerous positive outcomes in diverse sectors.

REFERENCES

Adly, A. S., Adly, A. S., & Adly, M. S. (2020). Approaches based on artificial intelligence and the internet of intelligent things to prevent the spread of COVID-19: Scoping review. *Journal of Medical Internet Research, 22*(8), e19104–e19104.

Adzaho, G. (2013, 19 December). Can technology help preserve elements of culture in the digital age? Diplomacy Education. Retrieved 17 September 2021 from www.diplomacy.edu/blog/can-technology-help-preserve-elements -culture-digital-age/

Afrin, T., & Yodo, N. (2020). A Survey of Road Traffic Congestion Measures towards a Sustainable and Resilient Transportation System. *Sustainability, 12*(11), 4660.

Alam, M., St-Hilaire, M., & Kunz, T. (2017). An optimal P2P energy trading model for smart homes in the smart grid. *Energy Efficiency, 10*(6), 1475–1493.

Albert, S. (2019, 21 April). Smart cities: The promises and failures of utopia technological planning. *The Conversation*. Retrieved 27 January 2022 from https://theconversation.com/smart-cities-the-promises-and-failures-of-utopian -technological-planning-114405

Alexander, C., Ishikawa, S., & Silverstein, M. (1977). *A Pattern Language*. Oxford University Press.

Allam, Z. (2018). On smart contracts and organisational performance: A review of smart contracts through the blockchain technology. *Review of Economic and Business Studies, 11*(2), 137–156.

Allam, Z. (2019a). Achieving neuroplasticity in artificial neural networks through smart cities. *Smart Cities, 2*(2), 118–134.

Allam, Z. (2019b). The emergence of anti-privacy and control at the nexus between the concepts of safe city and smart city. *Smart Cities, 2*(1), 96–105.

Allam, Z. (2019c). Enhancing renewable energy adoption in megacities through energy diversification, land fragmentation and fiscal mechanisms. *Sustainable Cities Society, 101841*.

Allam, Z. (2020a). Biometrics, privacy, safety, and resilience in future cities. In *Biotechnology and Future Cities* (pp. 69–87). Springer International Publishing.

Allam, Z. (2020b). On culture, technology and global cities. In *Cities and the Digital Revolution: Aligning Technology and Humanity* (pp. 107–124). Springer International Publishing.

Allam, Z. (2020c). Privatization and privacy in the digital city. In *Cities and the Digital Revolution: Aligning Technology and Humanity* (pp. 85–106). Springer International Publishing.

Allam, Z. (2020d). The rise of Singapore. In *Urban Governance and Smart City Planning*. Emerald Publishing.

Allam, Z. (2020e). Seeking liveability through the Singapore model. In *Urban Governance Smart City Planning* (pp. 45–76). Emerald Publishing.

Allam, Z. (2020f). Singapore's governance style and urban planning. In *Urban Governance and Smart City Planning* (pp. 27–43). Emerald Publishing.

Allam, Z. (2021). Big data, artificial intelligence and the rise of autonomous smart cities. In *The Rise of Autonomous Smart Cities* (pp. 7–30). Springer International Publishing.

Allam, Z., & Dhunny, A. Z. (2019). On big data, artificial intelligence and smart cities. *Cities, 89*, 80–91.

Allam, Z., & Jones, D. (2018). Promoting resilience, liveability and sustainability through landscape architectural design: A conceptual framework for Port Louis, Mauritius, a small island developing state. IFLA World Congress Singapore.

Allam, Z., & Jones, D. J. E. (2019a). Climate change and economic resilience through urban and cultural heritage: The case of emerging small island developing states economies. *Economies, 7*(2), 62.

Allam, Z., & Jones, D. S. (2019b). Attracting investment by introducing the city as a special economic zone: A perspective from Mauritius. *Urban Research and Practice*, 1–7.

Allam, Z., & Jones, D. S. (2019c). The potential of blockchain within air rights development as a prevention measure against urban sprawl. *Urban Science, 3*(1), 38.

Allam, Z., & Jones, D. S. (2020). On the Coronavirus (COVID-19) outbreak and the smart city network: Universal data sharing standards coupled with artificial intelligence (AI) to benefit urban health monitoring and management. *Healthcare, 8*(1), 46.

Allam, Z., & Jones, D. S. (2021). Future (post-COVID) digital, smart and sustainable cities in the wake of 6G: Digital twins, immersive realities and new urban economies. *Land Use Policy, 101*, 105201.

Allam, Z., & Newman, P. J. S. C. (2018). Economically incentivising smart urban regeneration. Case study of Port Louis, Mauritius. *Smart Cities, 1*(1), 53–74.

Allam, Z., Jones, D., & Thondoo, M. (2020). Economically incentivizing urban sustainability and resilience. In Z. Allam, D. Jones, & M. Thondoo (Eds), *Cities and Climate Change: Climate Policy, Economic Resilience and Urban Sustainability* (pp. 83–106). Springer International Publishing.

Appendino, F. (2017). Balancing heritage conservation and sustainable development: The case of Bordeaux. *IOP, 245*(062002), 1–11.

Aram, F., Higueras García, E., Solgi, E., & Mansournia, S. (2019). Urban green space cooling effect in cities. *Heliyon, 5*(4), e01339.

Atkinson, R. (2000). The hidden costs of gentrification: Displacement in central London. *Journal of Housing and the Built Environment, 15*(4), 307–326.

Balletti, C., & Ballarin, M. (2019). An application of integrated 3D technologies for replicas in cultural heritage. *ISPRS International Journal of Geo-Information, 8*(6). https://doi.org/10.3390/ijgi8060285

Bibri, S. E. (2021a). Data-driven smart sustainable cities of the future: An evidence synthesis approach to a comprehensive state-of-the-art literature review. *Sustainable Futures, 3*, 100047.

Bibri, S. E. (2021b). Data-driven smart sustainable cities of the future: Urban computing and intelligence for strategic, short-term, and joined-up planning. *Computational Urban Science, 1*(1), 8.

Biyik, C., Allam, Z., Pieri, G., Moroni, D., O'fraifer, M., O'connell, E., Olariu, S., & Khalid, M. (2021). Smart parking systems: Reviewing the literature, architecture and ways forward. *Smart Cities, 4*(2), 623–642.

Camagni, R., Gibelli, M. C., & Rigamonti, P. (2002). Urban mobility and urban form: The social and environmental costs of different patterns of urban expansion. *Ecological Economics, 40*(2), 199–216.

Cech, T. G., Spaulding, T. J., & Cazier, J. A. (2018). Data competence maturity: Developing data-driven decision making. *Journal of Research in Innovative Teaching and Learning, 11*(2), 139–158.

Chametzky, J. (1989). Beyond melting pots, cultural pluralism, ethnicity: Or, déjà vu all over again. *MELUS, 16*(4), 3–17.

Chatterton, I., & Hanway, C. E. (2021, 18 June). Smart Cities 4.0: Engaging Citizens; Transforming Urban Infrastructure through Infratech and Open Data. CommDev; International Finance Corporation. Retrieved 28 January 2022 from https://commdev.org/blogs/smart-cities-4-0-engaging-citizens-transforming-urban-infrastructure-through-infratech-and-open-data/

Coldwell, W. (2014, 19 June). Paris becomes first city to extend bike sharing scheme to children. *Guardian*. Retrieved 27 January 2022 from www .theguardian.com/travel/2014/jun/19/paris-city-bike-hire-cycling-children -family

Crinson, M. (2017). Singapore's moment: Critical regionalism, its colonial roots and profound aftermath. *Journal of Architecture, 22*(4), 689–709.

Dabeedooal, J. Y., Dindoyal, V., Allam, Z., & Jones, S. D. (2019). Smart tourism as a pillar for sustainable urban development: An alternate smart city strategy from Mauritius. *Smart Cities, 2*(2). https://doi.org/10.3390/smartcities2020011

De Cesari, C., & Dimova, R. (2019). Heritage, gentrification, participation: Remaking urban landscapes in the name of culture and historic preservation. *International Journal of Heritage Studies, 25*(9), 863–869.

De Guimarães, J. C. F., Severo, E. A., Felix Júnior, L. A., Da Costa, W. P. L. B., & Salmoria, F. T. (2020). Governance and quality of life in smart cities: Towards sustainable development goals. *Journal of Cleaner Production, 253*, 119926.

De Vos, J. (2020). The effect of COVID-19 and subsequent social distancing on travel behavior. *Transportation Research Interdisciplinary Perspectives, 5*, 100121.

Deutsche, W. (2021, 11 May). Coronavirus inspires cities to push climate-friendly mobility. DW News. Retrieved 13 April 2022 from www.dw.com/en/ coronavirus-inspires-cities-to-push-climate-friendly-mobility/a-53390186

Diener, A., & Mudu, P. (2021). How can vegetation protect us from air pollution? A critical review on green spaces' mitigation abilities for air-borne particles from a public health perspective – with implications for urban planning. *Science of the Total Environment, 796*, 148605.

Dixit, A., Kumar Chidambaram, R., & Allam, Z. (2021). Safety and risk analysis of autonomous vehicles using computer vision and neural networks. *Vehicles, 3*(3). https://doi.org/10.3390/vehicles3030036

Dobbs, R., Smit, S., Remes, J., Manyika, J., Roxburgh, C., & Restrepo, A. (2015). Urban world: Mapping the economic power of cities. Retrieved 1 February 2022 from www.mckinsey.com/~/media/McKinsey/Featured%20Insights/ Urbanization/Urban%20world/MGI_urban_world_mapping_economic _power_of_cities_full_report.pdf

Duany, A., Plater-Zyberk, E., & Speck, J. (2000). *Suburban Nation: The Rise of Sprawl and the Decline of the American Dream*. North Point.

Eireiner, A. V. (2021). Promises of urbanism: New Songdo City and the power of infrastructure. *Space and Culture*, 1–11.

Elmeadawy, S., & Shubair, R. M. (2019, 19–21 November). 6G wireless communications: Future technologies and research challenges. 2019 International Conference on Electrical and Computing Technologies and Applications, Ras Al Khaimah, United Arab Emirates.

Fogelberg, H. (1998). Electric car controversies. Chalmers University of Technology. Retrieved from www.osti.gov/etdeweb/servlets/purl/20026365

Fussell, S. (2018, 21 November). The city of the future is a data-collection machine. *The Atlantic*. Retrieved 28 January 2022 from www.theatlantic.com/ technology/archive/2018/11/google-sidewalk-labs/575551/

Galyean, C. (2015). Levittown: The imperfect rise of the American suburbs. *U.S. History Scene*. Retrieved 29 January 2022 from https://ushistoryscene.com/article/levittown/

Gascó-Hernandez, M. (2018). Building a smart city: Lessons from Barcelona. *Communications of the ACM*, *61*(4), 50–57.

Gill, S. S., Tuli, S., Xu, M., Singh, I., Singh, K. V., Lindsay, D., Tuli, S., Smirnova, D., Singh, M., Jain, U., Pervaiz, H., Sehgal, B., Kaila, S., Misra, S., Aslanpour, M. S., Mehta, H., Stankovski, V., & Garraghan, P. (2019). Transformative effects of IoT, blockchain and artificial intelligence on cloud computing: Evolution, vision, trends and open challenges. *Internet of Things*, *8*, 1–33.

Gonzalez, P. M. (2021, 28 January). The AI market in Japan: Spearheading industry innovation. *Tokyoesque*. Retrieved 29 January 2022 from https://tokyoesque.com/ai-market-in-japan/

Gulistan News. (2020, 20 April). CHILE: A drone delivers medicine to the elderly who are socially isolated to prevent contracting the coronavirus, Zapallar, Chile. Twitter.

Himelic, J., & Kreith, F. (2008). Potential benefits of plug-in hybrid electric vehicles for consumers and electric power utilities. Proceedings of the ASME 2008 International Mechanical Engineering Congress and Exposition, Boston, MA.

Hortas-Rico, M., & Solé-Ollé, A. J. U. s. (2010). Does urban sprawl increase the costs of providing local public services? Evidence from Spanish municipalities. *Urban Studies*, *47*(7), 1513–1540.

Iamsterdam. (2021, 5 November). AI in the Amsterdam area generates five-fold return on investment, research finds. Iamsteradam. Retrieved 29 January 2022 from www.iamsterdam.com/en/business/news-and-insights/news/2021/ai-in-the-amsterdam-area-generates-fivefold-return-on-investment-research-finds

Iberdrola. (2021). Blockchain technology at the service of urban management. Iberdrola. Retrieved 27 January 2022 from www.iberdrola.com/innovation/blockchain-for-smart-cities-urban-management#:~:text=ADVANTAGES%20OF%20BLOCKCHAIN%20FOR%20GOVERNING%20CITIES&text=Cities%20can%20interconnect%20using%20blockchain,their%20inhabitants%20in%20real%20time

IDC. (2020, 28 July). IoT growth demands rethink of long-term storage strategies. IDC. Retrieved 22 August 2021 from www.idc.com/getdoc.jsp?containerId=prAP46737220

International Monetary Fund. (2021, 26 October). Projected GDP ranking. IMF. Retrieved 29 January 2022 from https://statisticstimes.com/economy/projected-world-gdp-ranking.php

IRENA. (2017). *IRENA Cost and Competitiveness Indicators: Rooftop Solar PV*. International Renewable Energy Agency.

IRENA. (2019). *Future of Solar Photovoltaic: Deployment, Investment, Technology, Grid Integration and Socio-economic Aspects*. International Renewable Energy Agency. Retrieved 1 February 2022 from https://irena.org/-/media/Files/IRENA/Agency/Publication/2019/Nov/IRENA_Future_of_Solar_PV_2019.pdf

IRENA. (2020). *Innovation Landscape Brief: Peer-to-Peer Electricity Trading*. International Renewable Energy Agency. Retrieved 1 February 2022 from https://irena.org/-/media/Files/IRENA/Agency/Publication/2020/Jul/IRENA_Peer-to-peer_trading_2020.pdf?la=en&hash=D3E25A5BBA6FAC15B9C193F64CA3C8CBFE3F6F41

Ismagilova, E., Hughes, L., Rana, N. P., & Dwivedi, Y. K. (2020). Security, privacy and risks within smart cities: Literature review and development of a smart city interaction framework. *Information Systems Frontiers*. https://doi.org/10.1007/s10796-020-10044-1

Kim, H. W., Aaron McCarty, D., & Lee, J. (2020). Enhancing sustainable urban regeneration through smart technologies: An assessment of local urban regeneration strategic plans in Korea. *Sustainability*, *12*(17), 6868.

Kong, L., Huang, T., Zhu, Y., & Yu, S. (2020). Preface. In L. Kong, T. Huang, Y. Zhu, & S. Yu (Eds), *Big Data in Astronomy* (pp. xiii–xiv). Elsevier.

Mancebo, F. (2020). Smart city strategies: Time to involve people. Comparing Amsterdam, Barcelona and Paris. *Journal of Urbanism: International Research on Placemaking and Urban Sustainability*, *13*(2), 133–152.

Marshall, C. (2015, 28 April). Levittown, the prototypical American suburb – a history of cities in 50 buildings, day 25. *Guardian*. Retrieved 28 January 2022 from www.theguardian.com/cities/2015/apr/28/levittown-america-prototypical-suburb-history-cities

McCarthy, N. (2020, 10 March). Traffic congestion costs US cities billions of dollars every year. Forbes. Retrieved 7 August from www.forbes.com/sites/niallmccarthy/2020/03/10/traffic-congestion-costs-us-cities-billions-of-dollars-every-year-infographic/

McGrath, J. M. (2020, 8 May). The real reason Sidewalk Labs failed in Toronto. TVO. Retrieved 28 January 2022 from www.tvo.org/article/the-real-reason-sidewalk-labs-failed-in-toronto

Mercader-Moyano, P., & Serrano-Jiménez, A. (2021). Special issue: Urban and buildings regeneration strategy to climatic change mitigation, energy, and social poverty after a world health and economic global crisis. *Sustainability*, *13*(21), 11850.

Moe, R., & Wilkie, C. (1999). *Changing Places: Rebuilding Community in the Age of Sprawl*. Henry Holt and Company.

Newcombe, T. (2014, 21 April). Santander: The smartest smart city. *Governing*. Retrieved 28 January 2022 from www.governing.com/archive/gov-santander-spain-smart-city.html

Nguyen, D. C., Ding, M., Pathirana, P. N., Seneviratne, A., Li, J., Niyato, D., Dobre, O., & Poor, H. V. (2021). 6G Internet of Things: A comprehensive survey. *IEEE Internet of Things Journal*, 1.

Nikonova, A., & Biryukova, M. (2017). The role of digital technologies in the preservation of cultural heritage. *Muzeologia a Kulturne Dedicstvo*, *5*, 169–173.

Nižetić, S., Šolić, P., López-de-Ipiña González-de-Artaza, D., & Patrono, L. (2020). Internet of Things (IoT): Opportunities, issues and challenges towards a smart and sustainable future. *Journal of Cleaner Production*, *274*, 122877–122877.

NYC. (2020). Technology and media industry. Retrieved 29 January 2022 from www1.nyc.gov/site/internationalbusiness/industries/technology-and-media -industry.page#:~:text=New%20York%20City%20is%20the,second%20only %20to%20Silicon%20Valley

Okeafor, J.-P. (2020, 10 November). How many cars are there in the world? And which country has most cars? Naija Auto. Retrieved 28 January 2022 from https://naijauto.com/market-news/how-many-cars-are-there-in-the-world -7100

Pereira, G. V., Cunha, M. A., Lampoltshammer, T. J., Parycek, P., & Testa, M. G. (2017). Increasing collaboration and participation in smart city govern-ance: A cross-case analysis of smart city initiatives. *Journal of Information Technology for Development, 23*(3), 526–553.

Peters, M. A., & Besley, T. (2021). 5G transformational advanced wireless futures. *Educational Philosophy and Theory, 53*(9), 847–851.

Population U. (2022). Tokyo population. Population U. Retrieved 29 January 2022 from www.populationu.com/cities/tokyo-population#:~:text=Tokyo %20population%20in%202022%20is,in%20the%20city%20center%20itself

Rathore, M. M., Shah, S., Shukla, D., Bentafat, E., & Bakiras, S. (2021). The role of AI, machine learning, and big data in digital twinning: A systematic literature review, challenges, and opportunities. *IEEE Access*, 1–1.

Research and Markets. (2021, 5 May). Global DNA data storage markets and technologies report 2021. Cision PR Newswire. Retrieved 4 August 2021 from www.prnewswire.com/news-releases/global-dna-data-storage-markets-and -technologies-report-2021-review-the-dna-data-storage-technologies-patents -including-dna-read-and-write-technologies-301284365.html

Riganti, P. (2017). Smart cities and heritage conservation: Developing a smart heritage agenda for sustainable inclusive communities. *Archnet-Ijar: International Journal of Architectural Research, 11*(3), 16–27.

Ritchie, H., & Roser, M. (2021). Emissions by sector. Our World in Data. Retrieved 28 January 2022 from https://ourworldindata.org/emissions-by -sector

Salingaros, N. A. (2000). Complexity and urban coherence. *Journal of Urban Design, 5*, 291–316. http://zeta.math.utsa.edu/~yxk833/UrbanCoherence.html

See, A. v. (2021, 7 June). Amount of data created, consumed, and stored 2010–2025. Statista. Retrieved 26 January 2022 from www.statista.com/ statistics/871513/worldwide-data-created/#:~:text=Over%20the%20next %20five%20years,to%20more%20than%20180%20zettabytes

Shahab, S., & Allam, Z. (2019, 7 November). Reducing transaction costs of trad-able permit schemes using blockchain smart contracts. *Growth and Change*. https://doi.org/10.1111/grow.12342

Shrestha, A. K., Vassileva, J., & Deters, R. (2020). A blockchain platform for user data sharing ensuring user control and incentives. *Frontiers, 3*(48). https://doi.org/10.3389/fbloc.2020.497985

Taylor, M. (2019, 10 June). Singapore taps into innovation as water imports start drying up. Scroll. Retrieved 29 January 2022 from https://scroll.in/article/ 926270/singapore-taps-into-innovation-as-water-imports-dry-up

UN-Habitat. (2020). People-centered smart cities. UN-Habitat. Retrieved 28 January 2022 from https://unhabitat.org/programme/people-centered-smart -cities

UNEP. (2014). Emerging issues for small island developing states. UNEP. Retrieved 1 February 2022 from https://sustainabledevelopment.un.org/ content/documents/1693UNEP.pdf

UNEP. (2015). Goal 11: Sustainable cities and communities. United Nations. Retrieved 17 January 2020 from www.unenvironment.org/explore-topics/ sustainable-development-goals/why-do-sustainable-development-goals -matter/goal-11

United Nations. (2011). Transforming our world: The 2030 Agenda for Sustainable Development (A/RES/70/1). United Nations. Retrieved 1 February 2022 from https://sustainabledevelopment.un.org/content/documents/21252030 %20Agenda%20for%20Sustainable%20Development%20web.pdf

United Nations. (2017). New urban agenda. H. I. Secretariat. Retrieved 1 February 2022 from https://uploads.habitat3.org/hb3/NUA-English.pdf

United Nations. (2020). Cities and pollution. United Nations. Retrieved 1 November 2021 from www.un.org/en/climatechange/climate-solutions/cities -pollution

United Nations Framework Convention on Climate Change. (2015). *Paris Agreement*. UNFCCC. Retrieved 8 August 2021 from https://unfccc.int/sites/ default/files/english_paris_agreement.pdf

United States Environmental Protection Agency. (2020). Using green roofs to reduce heat islands. EPA. Retrieved 28 January 2022 from www.epa.gov/ heatislands/using-green-roofs-reduce-heat-islands

Vandy, K. (2020, 2 October). Coronavirus: How pandemic sparked European cycling revolution. BBC. Retrieved 6 January 2021 from www.bbc.com/news/ world-europe-54353914

Weiss, M., Zerfass, A., & Helmers, E. (2019). Fully electric and plug-in hybrid cars: An analysis of learning rates, user costs, and costs for mitigating CO_2 and air pollutant emissions. *Journal of Cleaner Production, 212*, 1478–1489.

World History Encyclopedia. (2021). City timeline. World History Encyclopedia. Retrieved 27 January 2022 from www.worldhistory.org/timeline/city/

World Meter. (2022, 29 January). Countries in the world by population (2022). WorldoMeter. Retrieved 29 January 2022 from www.worldometers.info/ world-population/population-by-country/

Zhang, C., Wu, J., Zhou, Y., Cheng, M., & Long, C. (2018). Peer-to-peer energy trading in a microgrid. *Applied Energy, 220*, 1–12.

Zvolska, L., Lehner, M., Voytenko Palgan, Y., Mont, O., & Plepys, A. (2019). Urban sharing in smart cities: The cases of Berlin and London. *Local Environment, 24*(7), 628–645.

3. Smart cities must be sustainable and inclusive cities

INTRODUCTION

Cities across the globe are the 'engines' for economic growth, social integrations and diversity, centres for education, advanced healthcare, entertainment and political activities (De Cesari & Dimova, 2019; De Guimarães et al., 2020; Florida, 2017). They are also home to a majority of the global population – currently hosting approximately over 55 per cent and, by 2050, will be home to approximately 68 per cent of the global population (World Bank, 2021a). However, the different roles they play attract a myriad of negative influences on areas like sustainability, inclusivity and liveability (Allam & Jones, 2018; Allam et al., 2020). This has been affirmed in different forums and global policy documents such as the Sustainable Development Goals (SDGs) (United Nations Development Programme, 2015), the New Urban Agenda (United Nations, 2017) and the 2030 Agenda for Sustainable Development (United Nations, 2011), that directly address the need for cities to be restructured in ways that advance sustainability, liveability, safety and inclusivity outcomes.

With respect to sustainability agendas, activities such as transportation, manufacturing, urban construction and waste management have been found to elicit negative outcomes. These are influenced by increasing urban populations as well as the economic pursuits of urban dwellers (Kuddus et al., 2020; Ribeiro et al., 2019). Essentially, most cities are reported to host a sizeable number of residents that could be argued to have higher disposable incomes relative to their rural counterparts. Thus, a majority of these have the capacity to regularly use private cars for their normal travel, install energy-intensive electronic products and continue consuming manufactured products. They also have the capacity to initiate complex construction projects that are resource intensive including in the energy sector. Consequently, this leads to an increased pressure on

the energy sector as the quest to satisfy the increasing demand intensi-
fies. In the course of such consumption, a high amount of diverse waste
materials are released into the environment, of which a majority have
negative impacts on the environment. For instance, the transport sector
is credited with approximately 72 per cent of total global emissions of
carbon dioxide, with over 80 per cent of those emissions being directly
related to cities (European Environment Agency, 2020), especially
between the 1970s and 2019 where a lot of activity in the transport sector
was recorded (Wang & Ge, 2019). In terms of municipal solid waste,
which mostly arise from increased household consumption of manufactured goods, it is reported that global cities currently generate over 2.01
billion tonnes, and by 2050, the waste will increase to approximately
3.40 billion tonnes (World Bank, 2021b). This will translate to more
than a 200 per cent increase in waste compared to the global population
increase that is anticipated to occur during the same period. This means
that waste production per capita will increase exponentially, as demand
and consumption of diverse products continue to increase.

Besides waste production, cities are now contributing significantly to
natural resource depletion as demand for products such as food supply,
energy, housing, infrastructure development, water, recreation facilities,
etc. continue to rise. While it is appreciated that cities occupy approximately 2 per cent of the earth's habitable surface only, it is noteworthy
that they account for more than 75 per cent of natural resources consumed globally (UNEP, 2022). This is alarming as only 55 per cent of the
world is living in cities, and with the projected urban population increase
trends, cities will endanger the sustainability of most of the earth's
natural resources. This could only be mitigated if urgent and proactive
approaches are adopted to find alternative options in different sectors
that are not resource intensive. For instance, in the energy sector, which
is a major resource-consuming industry as well as a main contributor to
global emissions, cities will need to urgently transition to use alternative
energy sources such as solar and wind (Zaheer, 2020).

The liveability aspect is a huge concern in cities, with numerous issues
such as climate change and congestion mostly in the road transport sector
and in informal settlements. The liveability agenda is further influenced
by the economic status of the residents. On this, cities have the greatest
levels of inequality, especially with regard to income (Allam, 2018;
Valcárcel-Aguiar et al., 2019). It is noted that since the 1980s, income
inequality and exclusion have continued to increase such that by 2020
over two-thirds of the global population can be described to have faced

some form of socioeconomic exclusion (UN-Habitat, 2020). This could be demonstrated by indicators such as unequal distribution of basic commodities such as affordable housing, clean water, education and healthcare. In the wake of the COVID-19 pandemic, those inequalities were greatly exposed, with a majority of the urban population, especially in low-income cadres, having no alternative in cases where they were required to isolate in order to prevent spreading the virus (Allam & Jones, 2020; Boza-Kiss et al., 2021). These diverse urban inequality challenges together have resulted in a decreased liveability index.

The inequalities in cities are further exposed by the surging number of homeless people, urban poor and increasing number of informal settlements across the globe. Regarding the urban poor, before the emergence of COVID-19, the numbers were approximately 27 million after a 9.2 per cent decline in poverty levels from the figures reported in 2017 (World Bank, 2020a). While it may be true that the number of extreme poor was declining, those numbers are still significant and call for extra attention. The situation worsened after the emergence of COVID-19, however, which is reported to have prompted an increase of people experiencing extreme poverty by a further 160 million people by the end of 2021 (Oxfam International, 2022). A majority of those are urban residents who in one way or another lost their livelihoods, their jobs and other such economic opportunities. The impacts of COVID-19 therefore exposed the vulnerability of urban societies to different externalities, hence the need to ensure cities endeavour to be inclusive. This argument is affirmed in a report that showed that during the pandemic, as a majority of people slid into poverty, a significant minority of people increased their wealth approximately 35-fold between 2020 and 2021 (Peterson-Withorn, 2021).

The urgent need to address the inclusivity and sustainability concerns in cities is critical and as such have the possibility to jeopardise and derail the important roles that cities play in boosting the socioeconomic agendas of different economies. As will be discussed in this chapter, addressing the inclusivity and sustainability agenda can be done by embracing different elements of the smart city concept, which have been tried and tested in different regions across the world. This has subsequently been approved as an important potential alternative to the conventional urban planning models that have prevailed since the first cities emerged (Allam, 2020). But, as will also be discussed, the smart city concept is a 'work in progress' that requires revisiting regularly to tighten areas that have so far not worked. Particularly, it will be important to address concerns on reducing the inequality gap both in terms of socioeconomic dimensions as well as in financing the projects. This argument is particularly relevant in geographies where there are limited financial capacities, especially in the global south. This chapter addresses the sustainability and inclusivity agendas in terms of intra-city relationships as well as external relationships between different elements in cities as individual entities.

URBAN SUSTAINABILITY

In the Paris Agreement that was officially adopted by over 196 parties on the 12 December 2015 during COP 21, the main objective was to ensure that parties agreed to reduce their carbon footprint (UNFCCC, 2015). Among the objectives was to persuade members to adopt socioeconomic practices that would ensure that global warming would be maintained below 2 degrees Celsius, and if possible, below 1.5 degrees Celsius – to pre-industrial levels (UNFCCC, 2021a). Since 4 November 2016 when the treaty came into force, numerous strides have been made toward its actualisation in different regions. While the document did not expressly mention cities as the main target areas to which much effort and concentration should be directed, reports from other sources confirm that urban areas are indeed the main parts of global economies on which maximum attention should be focused (Bulkeley & Castán Broto, 2013; OECD, 2014; Thaler et al., 2021). This is backed by reports that show that cities are responsible for approximately 70 per cent of the total emissions released into the atmosphere globally. This happens partly due to cities' robust development agendas that result in excessive resource consumption as well as increased waste production. These statistics have been reported despite the availability of post-2015 documents (Siemens, n.d.) such as the Paris Agreement, the SDGs and others that highlight the need for a sustainability agenda.

The calls for sustainability, however, are not only reported on global-scale policy documents but are also hinged on realities confronting different regions – more so urban areas. One reality is that most economic activities being undertaken in cities are resource intensive and have the potential to continue compromising the future availability of these resources (Allam & Newman, 2018b; George et al., 2015; Restrepo & Morales-Pinzón, 2018). For instance, the conventional energy production demand, mainly prompted by increasing urban populations and an influx of energy-intensive devices, is reported to be among the most unsustainable global practices. This is interesting as even some of the smart technologies championed to have the potential to solve urban challenges are pressuring the energy sector. For instance, most devices that are powered by Internet of Things (IoT) technologies have a high affinity for energy consumption, prompting scientists and other stakeholders to explore alternative ways of not only powering them but also storing the massive data they generate. The challenge of high-power consumption is compounded by the activities of many companies producing IoT products, where they consciously or subconsciously ensure that their products are uniquely designed. This uniqueness inhibits the standardisation of protocols and networks, hence adding to the numerous networks and central servers that are considered to be power intensive (THALES, 2021). Urban populations for their part add to a decline in resources, due to increased demand for housing, public infrastructures, manufactured goods, energy and waste disposal among other things.

Cities, however, should not succumb to those diverse realities, but as proposed in SDG 11, they should endeavour to become inclusive, safe, resilient, sustainable and conducive for human settlements. The smart city concept is being fronted as a potential candidate to solve diverse challenges. From pilot use cases in different regions, the model is considerably more efficient than previous planning models (Sharifi & Allam, 2021). For instance, in the urban sector, it has been highlighted in the 2021 report on SDG 11 that only half of the global urban population has convenient access to public transport that could be termed as efficient, safe and 'green' (United Nations Department of Economic and Social Affairs, 2021b). This has left a majority of urban residents having to rely on private cars that do not only increase congestion, but also add to the emission burden as well as increasing accident rates in urban areas (Frazer, 2019; Gössling, 2020; Sumantran et al., 2017). With potential smart technologies, it has been proven that it is now possible to adopt alternative mobility options such as ride sharing, car pooling, the use of bicycles and other options that not only increase convenience and reduce congestion, but also help save time and money and reduce emissions (Gambella et al., 2019; Shaaban, 2020). This further reduces the infrastructure development burden that most cities are facing due to the increased number of private cars. Infrastructure development, especially with regard to road expansion, is one initiative that increases the unsustainability activities where resources such as forests, arable lands, water bodies and game reserves are compromised to accommodate urban expansion needs (Alamgir et al., 2019; Philippe & Haigh, 2021). A case in point are the impacts of railway expansion on the Nairobi National Park, Kenya that included human–wildlife conflicts, disruption and alteration of the ecological setup of the park and increasing animal fatalities among other things (Ambani & Mulaku, 2021).

Cities' unsustainable practices have further been compounded in the energy sector, where by the end of 2020, fossil fuels were still the most popular resource used, accounting for over 80 per cent of world energy production (Abas Kalair et al., 2015). This is true despite the cost of producing energy from alternative sources becoming favourable in recent years, courtesy of technology use in the production of components such as solar panels, batteries and wind turbines (Dhunny et al., 2019a, 2019b, 2020). As of 2020, it is reported that the total alternative energy mix accounted for only 11.2 per cent, which is a slight improvement from the 8.7 per cent that was recorded in 2009 (Chestney, 2021).

Whereas there are numerous cases that could be cited to demonstrate that cities are yet to achieve the desired sustainability aspects in different frontiers, it has been noted that the idea of digitalisation is helping mitigate what would translate to future calamities. For instance, the spirited effort in the adoption and implementation of smart homes initiatives has been observed to have the potential to reduce carbon footprints from households by approximately 12.78 per cent (Gandzeichuk, 2021) – a significant achievement in view of the increasing population (Bhati et al., 2017). Smart homes are not only based on energy consumption but also on low-cost solutions, including the consumption of natural resources during construction. The aspect of automation that is synonymous with most smart homes allows for energy efficiency, increased comfort and liveability status and a high level of environmental sustainability considerations. When these are coupled with other urban digital solutions like smart mobility, smart energy systems and smart citizens, they have the capacity to allow for sustainability agendas to be pursued effectively and robustly (Sharifi & Allam, 2021). For instance, when all these aspects are coupled with characteristics such as proximity, density, diversity and ubiquitousness, championed in the 15-minute city concept proposed by Carlos Moreno (2017), it is possible to enhance the sustainability agendas in cities, especially post-COVID-19, which has prompted paradigm shifts in most global frontiers (Deloitte, 2020).

ELUSIVE INCLUSIVITY IN CITIES

Cities across the globe have been labelled as centres for diversity in terms of culture, socioeconomic status, education, economy and many other aspects (Colenbrander, 2016; Hatti & Rauhut, 2017; Jänicke,

2012). These make cities vibrant and attractive to a wider demographic composition, hence justifying the current trends of rapid urban population growth (Silva et al., 2018). However, such positive outcomes only represent the face value of most cities. When the surface is scratched, a plethora of issues relating to how urban residents interact emerges, with the most dominant characteristics being social and economic inequality. In the melting pot theory (Maddern, 2013), the emphasis is that, as people from different backgrounds continue to live together, there are possibilities that new cultures could emerge and these would somehow help overcome some of the shortcomings prominent in different cultures. However, while it may be an underestimation that such cultures do exist, it may be safe to argue that even new cultures do not solve most urban fragmentation challenges, but instead somehow strengthen them. This assertion can be explained by analysing the social, economic and political divisions that exist in most cities.

On the social front, it has been noted that cities represent the clearest picture of how socially fragmented the world population is (Nijman & Wei, 2020; Pandey, 2016; Shutters et al., 2021). The highest number of billionaires live in major cities, especially in the most developed and developing countries. Likewise, cities are also home to most of the poorest demographic (Kuddus et al., 2020). On this, it is noted that of the 4.4 billion people who reside in cities, one-quarter (1 billion) are located in slums and informal settlements and are subjected to abject poverty and deprivation (Khan, 2020). Across the board they experience unprecedented difficulties related to an acute shortage of clean water, proper sanitation and drainage, quality healthcare and education, waste management programmes, security of land tenure, scarcity of green spaces and recreation parks, etc. For those living outside of slums and informal settlements, a sizeable number have been rendered homeless due to erratic real estate activities, rising costs of living prompted by the slumping of global economies and cases of gentrification prompted by the adoption of technologies among other issues. On homelessness, it is reported that by the end of 2019 approximately 2 per cent (equivalent to 150 million people) from different urban areas were totally homeless, and another 1.6 billion of the urban population were experiencing challenges of getting proper shelter due to inadequate housing (Chamie, 2017). Those challenges were happening even during the period in which the global real estate industry was reported to have continued sustained growth. For instance, between 2018 and 2019, it is reported that the market grew at a compounded annual growth rate of approximately 4.1 per cent to reach a high of $9.6 trillion from $8.9 trillion (the growth rate in this period was 7.8 per cent) (MSCI, 2020).

The elusive inclusivity in cities is prompted by a number of factors, including the capitalist approach to urban management, increasing gentrification and political activities that influence urban planning policies. Regarding capitalism, it is evident that most factors of production in cities – e.g. the service industry, manufacturing, transport, entertainment industry and sport industry – are owned by a small group of the urban population (the elites). Karl Marx noted that in such cases, most of the urban masses are alienated in terms of income capping and market forces (Marx & Engels, 1885). The forces have made it possible for global markets for housing, entertainment, healthcare, education and others to continue growing, but only benefitting a few of urban residents. Such alienations are further accelerated by modern urban practices like gentrification (Atkinson, 2000), mostly driven by the availability of technologies, including smart city technologies. When this happens, minority groups, poor people, original home owners, migrants, minority religious groups and others are 'unconsciously' evicted and forced to seek alternative places to live. Those that are left are forced to put up with a high cost of living (food, digital products like the internet and others) and social alienation as their cultural background is disrupted as a majority of their peers migrate to other locations (Rodríguez-Pose & von Berlepsch, 2019).

The lack of inclusivity in cities is further compounded during decision making and policy formulation. On this, one selling point for the smart city concept is the aspect of enhanced public participation through diverse platforms such as social media (Batty et al., 2012; Sitinjak et al., 2018). However, in most cases, the aspirations, decisions and demands of the most vulnerable groups are left out or are not given priority during the design and implementation stages in many urban projects (Arnstein, 1969; Cardullo & Kitchin, 2019). This explains the unfortunate scenarios and circumstances that are prompted by gentrification. But a lack of participation happens amid the presence of policy documents such as SDG 11, advocating for inclusivity in all aspects of city management which in this case would include public participation to facilitate bottom-up approaches in decision making rather than the dominant trickle-down approach. The end results of implementing programmes and projects designed at the top without involving the general public is the creation of an urban environment that does not solve the inequality gap, but on the contrary helps it to continue growing.

The different smart technologies, however, could be used to help solve some of the social and economic inequalities and inclusivity shortcomings in cities. For instance, in the production and distribution of services such as energy, data generated by different smart devices could help in decision making, identifying priority locations and alternative methods that could be used to generate more energy to meet increasing demand (Allam & Dhunny, 2019). Further, smart technologies could help local governments promote the use of alternative mobility methods, including cycling and walking, hence reducing the exclusion propagated by private car ownership (Frazer, 2019; Stamatiadis et al., 2017). In the housing sector, smart technologies can help in the adoption of mixed-use housing programmes, thus making neighbourhoods safer, increasing economic opportunities, promoting the availability of recreation centres and other things that get little attention in the case of conventional planning models that emphasise vehicular flows rather than human-scale dimensions.

WHY THE SMART CITY MUST ADDRESS URBAN SUSTAINABILITY AND INCLUSIVITY

In the historical accounts of cities, from 6500 BCE (World History Encyclopedia, 2021) to current accounts, the aspect of inclusivity has never been achieved. There has been persistent and visible demarcation between the rich, the middle class and the poor in terms of social, economic and spatial inclusion. However, while the inclusivity debate has continued as cities continue to transform and new ones emerge, the debate on sustainability officially began in 1987, where a universally agreed definition on the same was provided in the Brundtland Report (Keeble, 1988). The debate began when economies, especially those in Western societies, began to realise that their pursuit of diverse economic and industrial activities had notable impacts on their environment and social balance. Since then, a spirited and concerted attention was focused on ensuring that all economic and development activities are pursued in a way that encourages sustainability. Although the aspect of sustainability only began gaining traction when modern cities emerged, and urban managers started to explore how they could integrate technologies (mostly through the use of data) to enhance different agendas (Alberts et al., 2017; Brasuell, 2015), it has become a buzz word in the planning arena (Allam, 2018a; Allam & Newman, 2018b). These issues have been identified as among modern-day urban challenges that need to be addressed amicably and in a concerted manner if cities are to continue sustaining their economic, social and political roles in different countries.

The SDGs make explicitly clear the need for sustainability and inclusivity. In SDG 11, the attention is exclusively focused on the need for cities to be sustainable, safe and resilient (United Nations Department of Economic and Social Affairs, 2021b). While the SDGs were formulated in 2015 after lengthy discussions and debates, the content and goals therein have been vindicated by the occurrences that have befallen cities and human settlements. Among such occurrences is the increase in climate change impacts that have not only affected vulnerable regions (small island developing states, coastal cities and low-lying regions) (Mycoo & Donovan, 2017), but have also significantly affected diverse regions in developed countries (Costa & Floater, 2015). For instance, as of 2020, most cities in Europe and North America had experienced unprecedented challenges of flooding and erratic heavy precipitation (Alipour et al., 2020; Henson & Masters, 2021). These regions have also been exposed to heatwaves, droughts, wildfires and vector-borne diseases (European Environment Agency, 2021). It was predicted in 2018 that approximately 251 cities across Europe would experience very severe to extreme weather conditions (Schlanger, 2018) and this prediction has materialised. For instance, in 2021 alone, most cities in Europe, especially in Germany and Belgium, are reported to have experienced catastrophic flooding, with over 125 fatalities reported (Erdbrink, 2021). Further, during the same period, it was reported that drought incidences had increased in the region and serious outcomes would be experienced in the coming years – with an increase of over 50 per cent in some cities in France (Carrington, 2021).

The impacts of climate change, however, are expected to be widespread in different cities across the globe regardless of their geographical profile. This means that cities in already identified vulnerable regions like small island developing states and coastal regions are expected to continue paying heavily even though their carbon footprint is almost negligible (less than 1 per cent of total global emissions) (UNEP, 2014; UNFCCC, n.d.). In the latest report by the United Nations Framework Convention on Climate that preceded the COP 26 meeting that took place in Glasgow (UNFCCC, 2021b), it was feared that the global temperatures would exceed 2 degrees Celsius (UNFCCC, 2021a), meaning global societies and economies need to brace for extreme weather and associated climate change crises.

While there is no alternative for global economies, especially cities, than to increase their attention on sustainability efforts, including embracing smart technologies that guarantee some positive outcomes in the end, the aspect of inclusivity needs to be addressed swiftly. This was accentuated by the recent outbreak of the COVID-19 pandemic that exposed the social and economic tenets of most cities (Allam & Jones, 2021; Sharifi & Khavarian-Garmsir, 2020; United Nations, 2021; World Bank, 2020a; Zarrilli & Aydiner-Avsar, 2020). Vulnerable demographics like women, children and minority groups have experienced unprecedented challenges during the pandemic. For instance, in terms of economic inclusion, a majority of women in low-income economies and developing economies, especially in cities, lost their livelihoods in 2020 during the height of COVID-19 (Zarrilli & Aydiner-Avsar, 2020). This was attributed to the fact that most women are employed in informal sectors and most of those were disrupted by the health measures of different countries. However, it is important to explore why most women are employed in informal sectors (UN Chronicle, 2021; Zarrilli & Aydiner-Avsar, 2020). SDG 5 advocated for gender equality and empowerment of women and girls, especially in crucial decision making, and this has not been achieved in most cities (United Nations, 2015). As of 2020, it was reported that only 36.3 per cent of women are part of local government leadership positions, hence rendering their contribution nominal when compared to their male counterparts (United Nations Department of Economic and Social Affairs, 2021a). The aspect of social inclusion in cities does not only entail women and girls, but also needs to incorporate other minority groups. Therefore, as cities undertake different digital solutions, it will be paramount to ensure that this assures the achievement of inclusivity and sustainability agendas. These should include regeneration and conservation programmes in sectors like infrastructure, housing, culture and others need not disadvantage various groups as is the case of gentrification experienced in most cities.

Pursuing sustainability and inclusivity agendas in cities has the potential to allow cities to achieve major milestones in economic development, with housing, healthcare and education for the majority of the population. However, when these are neglected, it will render initiatives like the digitalisation of cities null and void as they need to ensure cities are conducive for human settlement. This cannot be attained unless there is parity in different sectors.

CHALLENGES THAT MUST BE OVERCOME TO ACHIEVE THE OBJECTIVE

A majority of global cities are reported to have fallen short in terms of pursuing inclusivity and sustainability. Such trends have rendered the achievement of global agendas such as the SDGs, the Paris Agreement, the New Urban Agenda and the 2030 Agenda for Sustainable Development to be a mirage. However, it has been observed that most cities have formulated relevant policies and set prerequisite objectives toward becoming inclusive as well as incorporating sustainable development agendas in their planning pursuits (United Nations Department of Economic and Social Affairs, 2021b). This is affirmed in a 2021 SDG progress report that showed that approximately 156 countries across the globe have developed national urban policies. Unfortunately, only half of those have taken initiatives to implement the policies (United Nations Department of Economic and Social Affairs, 2021a). It is important to understand that cities, which are the main target for such policies, experience varied challenges that when compounded make it hard to balance the elements of inclusivity as well as sustainability. The challenges include insufficient financial resources, lack of cooperation between different stakeholders (e.g. governments at different levels), relying on unsustainable planning models, limited institutional capacities, etc. (Allam et al., 2021; Baguant-Moonshiram et al., 2013; Bloomfield & Steward, 2020; Cullen-Knox et al., 2017).

The availability of financial resources and their subsequent distribution and utilisation is a major bane for many cities in their transformation agendas, especially in the pursuit of sustainability and inclusivity. In a report by Deloitte (2013), it was highlighted that very few cities across the globe have the capacity to effectively finance any form of urban transformation or investment programme from their public financial resources. A majority of urban areas, especially in developing and less developed economies, have to seek alternative financial sources to support their investment agendas. The smart cities planning model, if supported by appropriate fiscal mechanisms, can be a potent solution to help solve the twin challenges of sustainability and inclusivity in cities (Allam & Newman, 2018a). However, this model has also been flagged as capital intensive, hence most cities would require external financial support from institutions such as banks (Bossone & Sarr, 2016). It has been observed that most of the financial resources from external sources attract huge interest rates and numerous pre-qualification requirements that discourage many cities from investing in substantial transformational projects (Deloitte, 2013). Further, most economies have already secured loans from institutions such as the International Monetary Fund, with a number of those countries plunging into huge debt that would automatically disqualify their cities from securing loans (Kose et al., 2020). For instance, during the height of the COVID-19 pandemic (between 2020 and 2021), it is reported that most developing and least developed economies incurred excessive debt (International Monetary Fund, 2021; Kling et al., 2018; Nishio, 2021). Some countries had to extend their debt ceilings to qualify for more debt quotas (UNCTAD, 2021). As such, some have been forced to restructure their economic model so as to manage to slowly service and repay both existing and new loans. This approach has overburdened several economies, which are already fragile, such that no substantial investment projects can be successfully implemented.

Another challenge prominent while financing projects earmarked for sustainability and inclusivity is the engagement of the general public in the critical decision-making process (Kateja, 2012; Tan & Taeihagh, 2020). As a result, some projects initiated in cities, despite the goodwill they attract, sometimes fail to appeal to locals, who end up rejecting them. This results in the emergence of ghost projects. The major stakeholders in urban planning include local governments, the private sector, residents and other participants who have to be represented in policy-making processes. However, in some cases, national governments overstep their mandate and propose investment programmes to be implemented in cities without sufficient input of local government or from the residents. Such trends have negative effects, including disharmony between different levels of governance and financial flow challenges, thus leading to the lagging or failing of projects that would ultimately promote sustainability and inclusivity. On this, the best decision-making channel that could be adopted to ensure projects initiated in cities guarantee sustainability development and promote social inclusion would be embracing bottom-up approaches (Arnstein, 1969; Guo, 2018). This would particularly benefit programmes such as urban rejuvenation and regeneration programmes that are argued to result in gentrification rather than create opportunities for social vibrancy and economic growth. Such models need to be contextualised to fit the sociopolitical fabric in question. A completely different model is that of Singapore, in view of its autocratic governance style, but which successfully generated economic and technological reforms (Allam & Allam, 2020a, 2020b, 2020c; Allam & Jones, 2018).

Sustainability and inclusivity aspects could be derailed in urban areas by local governments relying on unsustainable planning models. For instance, before the emergence of the smart city planning approach, most cities were fashioned to promote vehicular flow and modernists' aesthetic pursuits without giving priority to human dimensions (Banister, 2012; Jacobs, 1961). As a result, most cities continue to struggle with challenges such as traffic congestion, low housing and insufficient infrastructure prompted by an unprecedented increase in the number of vehicles and other challenges. It is relatively expensive to remodel cities to incorporate elements such as bicycle lanes, green spaces and affordable housing programmes that would promote sustainable practices as well as inclusivity (Alamgir et al., 2019; APMG International, 2016; Atkinson, 2000; Camagni et al., 2002). The challenges of relying on planning models that give little attention to human aspects were exposed in most cities during the height of the COVID-19 pandemic, where over half of the world was 'trapped' in lockdowns (Boza-Kiss et al., 2021; El Said, 2021; International Trade Centre, 2020). As a result, most of the global gross domestic product plummeted to approximately 3.5 per cent, with that of some individual countries experiencing even higher declines (World Bank, 2020b). After countries started to reopen their economies, massive incidences of unsustainable practices were reported, with most of these happening in cities.

The smart city concept has the potential to allow most cities to fast-track their commitment toward achieving sustainable development agendas as well as address the different elements of inequalities that exist. However, this will need to be backed by genuine cooperation between different stakeholders, from decision making to financing, while paying attention to the pertinent needs of the residents. Priority will need to be focused on projects that not only guarantee sustainable development but also promote equity, for instance investing in projects such as alternative energy that would ensure reliable connectivity as well as reduced emissions. Achieving these urban elements will be made possible by ensuring challenges such as financing, non-cooperation, the digital divide and others are overcome, especially allowing vulnerable economies and those financially constrained to initiate programmes that have the capacity to address both sustainability and inclusivity challenges.

REFERENCES

Abas Kalair, N., Kalair, A., & Khan, N. (2015). Review of fossil fuels and future energy technologies. *Futures, 69.* https://doi.org/10.1016/j.futures.2015.03.003

Alamgir, M., Sloan, S., Campbell, M. J., Engert, J., Kiele, R., Porolak, G., Mutton, T., Brenier, A., Ibisch, P. L., & Laurance, W. F. (2019). Infrastructure expansion challenges sustainable development in Papua New Guinea. *PLOS ONE, 14*(7), e0219408.

Alberts, G., Went, M., & Jansma, R. (2017). Archaeology of the Amsterdam digital city: Why digital data are dynamic and should be treated accordingly. *Internet Histories, 1*(1–2), 146–159.

Alipour, A., Ahmadalipour, A., & Moradkhani, H. (2020). Assessing flash flood hazard and damages in the southeast United States. *Journal of Flood Risk Management, 13*(2), e12605.

Allam, M. Z. (2018). *Redefining the Smart City: Culture, Metabolism and Governance: Case Study of Port Louis, Mauritius.* Curtin University. https://espace.curtin.edu.au/handle/20.500.11937/70707

Allam, Z. (2018). Contextualising the smart city for sustainability and inclusivity. *New Design Ideas, 2*(2), 124–127.

Allam, Z. (2020). *Urban governance and smart city planning: Lessons from Singapore.* Emerald Group Publishing.

Allam, Z., & Allam, Z. (2020a). The rise of Singapore. In *Urban Governance and Smart City Planning* (pp. 1–26). Emerald Publishing.

Allam, Z., & Allam, Z. (2020b). Seeking liveability through the Singapore model. In *Urban Governance and Smart City Planning* (pp. 45–76). Emerald Publishing.

Allam, Z., & Allam, Z. (2020c). Singapore's governance style and urban planning. In *Urban Governance and Smart City Planning* (pp. 27–43). Emerald Publishing.

Allam, Z., & Dhunny, Z. A. (2019). On big data, artificial intelligence and smart cities. *Cities, 89,* 80–91.

Allam, Z., & Jones, D. (2018). Promoting resilience, liveability and sustainability through landscape architectural design: A conceptual framework for Port Louis, Mauritius; a small island developing state. IFLA World Congress Singapore.

Allam, Z., & Jones, D. S. (2020). Pandemic stricken cities on lockdown. Where are our planning and design professionals (now, then and into the future)? *Land Use Policy, 97,* 104805.

Allam, Z., & Jones, D. S. (2021). Future (post-COVID) digital, smart and sustainable cities in the wake of 6G: Digital twins, immersive realities and new urban economies. *Land Use Policy, 101,* 105201.

Allam, Z., & Newman, P. (2018a). Economically incentivising smart urban regeneration: Case study of Port Louis, Mauritius. *Smart Cities, 1*(1), 53–74.

Allam, Z., & Newman, P. (2018b). Redefining the smart city: Culture, metabolism and governance. *Smart Cities, 1*(1), 4–25.

Allam, Z., Jones, D., & Thondoo, M. (2020). Climate change mitigation and urban liveability. In *Cities and Climate Change* (pp. 55–81). Palgrave Macmillan.

Allam, Z., Sharifi, A., Giurco, D., & Sharpe, S. A. (2021). On the theoretical conceptualisations, knowledge structures and trends of green new deals. *Sustainability, 13*(22), 12529.

Ambani, M. M., & Mulaku, G. C. J. J. o. E. P. (2021). GIS assessment of environmental footprints of the Standard Gauge Railway (SGR) on Nairobi National Park, Kenya. *Journal of Environmental Protection, 12*(10), 694–716.

APMG International. (2016). How a private finance PPP project is financed: Where the money to pay construction costs comes from. In *PPP Certification Guide*. World Bank Group.

Arnstein, S. R. (1969). A ladder of citizen participation. *Journal of the American Institute of Planners, 35*(4), 216–224.

Atkinson, R. (2000). The hidden costs of gentrification: Displacement in central London. *Journal of Housing and the Built Environment, 15*(4), 307–326.

Baguant-Moonshiram, Y., Samy, M., & Thomas, K. (2013). *The Challenges of Building Sustainable Cities: A Case Study of Mauritius*. WIT Press.

Banister, D. (2012). Assessing the reality: Transport and land use planning to achieve sustainability. *Journal of Transport and Land Use, 5*(3), 1–14.

Batty, M., Axhausen, K. W., Giannotti, F., Pozdnoukhov, A., Bazzani, A., Wachowicz, M., Ouzounis, G., & Portugali, Y. (2012). Smart cities of the future. *European Physical Journal Special Topics, 214*(1), 481–518.

Bhati, A., Hansen, M., & Chan, C. M. (2017). Energy conservation through smart homes in a smart city: A lesson for Singapore households. *Energy Policy, 104,* 230–239.

Bloomfield, J., & Steward, F. (2020). The politics of the Green New Deal. *The Political Quarterly, 91*(4), 770–779.

Bossone, B., & Sarr, A. (2016, 18 February). Leveraging urbanisation to fund sustainable development and financial inclusion. World Bank. Retrieved 5 February 2022 from https://blogs.worldbank.org/allaboutfinance/leveraging -urbanization-fund-sustainable-development-and-financial-inclusion

Boza-Kiss, B., Pachauri, S., & Zimm, C. (2021). Deprivations and inequities in cities viewed through a pandemic lens. *Frontiers, 3.* https://doi.org/10.3389/ frsc.2021.645914

Brasuell, J. (2015, 22 June). The early history of the 'smart cities' movement – in 1974 Los Angeles. Planetizen. Retrieved 20 January 2021 from www .planetizen.com/node/78847

Bulkeley, H., & Castán Broto, V. (2013). Government by experiment? Global cities and the governing of climate change. *Transactions of the Institute of British Geographers, 38*(3), 361–375.

Camagni, R., Gibelli, M. C., & Rigamonti, P. (2002). Urban mobility and urban form: The social and environmental costs of different patterns of urban expansion. *Ecological Economics, 40*(2), 199–216.

Cardullo, P., & Kitchin, R. (2019). Being a 'citizen' in the smart city: Up and down the scaffold of smart citizen participation in Dublin, Ireland. *GeoJournal, 84*(1), 1–13.

Carrington, D. (2021, 15 March). Climate crisis: Recent European droughts 'worst in 2,000 years'. *Guardian*. Retrieved 4 February 2022 from www.theguardian .com/environment/2021/mar/15/climate-crisis-recent-european-droughts-worst-in -2000-years

Chamie, J. (2017, 13 July). As cities grow, so do the numbers of homeless. Yale Global. Retrieved 3 February 2022 from https://archive-yaleglobal.yale.edu/ content/cities-grow-so-do-numbers-homeless

Chestney, N. (2021, 15 June). Global fossil fuel use similar to decade ago in energy mix, report says. Reuters. Retrieved 3 February 2022 from www.reuters .com/business/environment/global-fossil-fuel-use-similar-decade-ago-energy -mix-report-says-2021-06-14/#:~:text=REN21%20said%20the%20share%20of ,in%202009%2C%20the%20report%20said

Colenbrander, S. (2016). *Cities as Engines of Economic Growth: The Case for Providing Basic Infrastructure and Services in Urban Areas*. IIED.

Costa, H., & Floater, G. (2015). Economic costs of heat and flooding in cities: Cost and economic data for the European Clearinghouse databases. *Reconciling Adaptation, Mitigation and Sustainable Development of Cities*. https://climate-adapt.eea.europa.eu/metadata/publications/economic-costs-of -heat-and-flooding-in-cities/ramses_2015_economic-costs-of-climate-change -in-european-cities.pdf

Cullen-Knox, C., Eccleston, R., Haward, M., Lester, E., & Vince, J. (2017). Contemporary challenges in environmental governance: Technology, governance and the social licence. *Environmental Policy and Governance*, *27*(1), 3–13.

De Cesari, C., & Dimova, R. (2019). Heritage, gentrification, participation: Remaking urban landscapes in the name of culture and historic preservation. *International Journal of Heritage Studies*, *25*(9), 863–869.

De Guimarães, J. C. F., Severo, E. A., Felix Júnior, L. A., Da Costa, W. P. L. B., & Salmoria, F. T. (2020). Governance and quality of life in smart cities: Towards sustainable development goals. *Journal of Cleaner Production*, *253*, 119926.

Deloitte. (2013). Funding options: Alternative financing for infrastructure development. Deloitte. Retrieved 1 February 2022 from www2.deloitte.com/ content/dam/Deloitte/au/Documents/public-sector/deloitte-au-ps-funding -options-alternative-financing-infrastructure-development-170914.pdf

Deloitte. (2020). Smart cities acing COVID-19 response using smart technology and data governance solutions. Deloitte. Retrieved 4 February 2022 from www2 .deloitte.com/content/dam/Deloitte/in/Documents/about-deloitte/in-about-deloitte -WEF-report-noexp.pdf

Dhunny, A. Z., Doorga, J. R. S., Allam, Z., Lollchund, M. R., & Boojhawon, R. (2019a). Identification of optimal wind, solar and hybrid wind-solar farming sites using fuzzy logic modelling. *Energy*, *188*, 116056.

Dhunny, A. Z., Allam, Z., Lobine, D., & Lollchund, M. R. (2019b). Sustainable renewable energy planning and wind farming optimization from a biodiversity perspective. *Energy*, *185*, 1282–1297.

Dhunny, A. Z., Timmons, D. S., Allam, Z., Lollchund, M. R., & Cunden, T. S. M. (2020). An economic assessment of near-shore wind farm development using a weather research forecast-based genetic algorithm model. *Energy*, 117541. https://doi.org/https://doi.org/10.1016/j.energy.2020.117541

El Said, G. R. (2021). How did the COVID-19 pandemic affect higher education learning experience? An empirical investigation of learners' academic performance at a university in a developing country. *Advances in Human-Computer Interaction*, 6649524. https://doi.org/10.1155/2021/6649524

Erdbrink, T. (2021, 16 July). Europe flooding deaths pass 125, and scientists see fingerprints of climate change. *New York Times*. Retrieved 4 February 2022 from www.nytimes.com/live/2021/07/16/world/europe-flooding-germany

European Environment Agency. (2020, 18 December). Greenhouse gases emission from transport in Europe. EEA. Retrieved 23 September 2021 from www.eea.europa.eu/data-and-maps/indicators/transport-emissions-of-greenhouse-gases-7/assessment

European Environment Agency. (2021, 18 January). Countries and cities in Europe urgently need to step up adaptation to climate change impacts. European Environment Agency. Retrieved 4 February 2022 from www.eea.europa.eu/highlights/countries-and-cities-in-europe#:~:text=The%20most%20pronounced%20impacts%20of,are%20also%20on%20the%20rise

Florida, R. (2017, 16 March). The economic power of cities compared to nations. CityLab. Retrieved 14 January 2020 from www.citylab.com/life/2017/03/the-economic-power-of-global-cities-compared-to-nations/519294/

Frazer, J. (2019, 6 August). The reshaping of city cores that were designed for cars. Forbes. Retrieved 5 December 2020 from www.forbes.com/sites/johnfrazer1/2019/08/06/the-reshaping-of-city-cores-that-were-designed-for-cars/?sh=55a97e261e46

Gambella, C., Monteil, J., Dekusar, A., Cabrero Barros, S., Simonetto, A., & Lassoued, Y. (2019). A city-scale IoT-enabled ridesharing platform. *Transportation Letters*, 1–7. https://doi.org/10.1080/19427867.2019.1694206

Gandzeichuk, I. (2021, 10 December). Smart homes: Reducing carbon footprint and expenses. Forbes. Retrieved 4 February 2022 from www.forbes.com/sites/forbestechcouncil/2021/12/10/smart-homes-reducing-carbon-footprint-and-expenses/?sh=72d0f5fd94b9

George, G., Schillebeeckx, S. D., & Liak, T. L. (2015). The management of natural resources: An overview and research agenda. *Academy of Management Journal*, *58*(6), 1595–1613.

Gössling, S. (2020). Why cities need to take road space from cars – and how this could be done. *Journal of Urban Design*, *25*(4), 443–448.

Guo, R. (2018). Citizen participation for inclusive outcomes: It takes a village to plan a city. *Urbans Solutions*, *12*, 54–63.

Hatti, N., & Rauhut, D. (2017). Cities and economic growth: A review. *Social Science Spectrum*, *3*(1), 1–15.

Henson, B., & Masters, J. (2021, 21 July). Central Europe staggers toward recovery from catastrophic flooding: More than 200 killed. Yale Climate Connections. Retrieved 6 November 2021 from https://yaleclimateconnections.org/2021/07/central-europe-staggers-toward-recovery-from-catastrophic-flooding-more-than-200-killed/

International Monetary Fund (2021, 16 March). IMF financing and debt service relief. IMF. Retrieved 30 March 2021 from www.imf.org/en/Topics/imf-and -covid19/COVID-Lending-Tracker

International Trade Centre. (2020). *COVID-19:* The great lockdown and its impact on small business. International Trade Centre. Retrieved 1 February 2022 from www.intracen.org/uploadedFiles/intracenorg/Content/Publications/ ITCSMECO2020.pdf

Jacobs, J. (1961). *The Death and Life of Great American Cities*. Vintage Books.

Jänicke, M. (2012). 'Green growth': From a growing eco-industry to economic sustainability. *Energy Policy*, *48*, 13–21.

Kateja, A. (2012). Building infrastructure: Private participation in emerging economies. *Procedia – Social and Behavioral Sciences*, *37*, 368–378.

Keeble, B. R. (1988). The Brundtland report: 'Our common future'. *Medicine and War*, *4*(1), 17–25.

Khan, A. (2020, 7 February). Cities of missed opportunities: Improving the lives of people in poverty. ODI. Retrieved 3 February 2022 from https://odi.org/en/ insights/cities-of-missed-opportunities-improving-the-lives-of-people-in-poverty/

Kling, G., Lo, Y., Murinde, V., & Volz, U. (2018). Climate vulnerability and the cost of debt. *SSRN Electronic Journal*, 1–30.

Kose, M. A., Ohnsorge, F., Nagle, P., & Sugawara, N. (2020). Caught by the cresting debt wave. *Finance and Development*, *57*(2), 40–43.

Kuddus, M. A., Tynan, E., & McBryde, E. (2020). Urbanization: a problem for the rich and the poor? *Public Health Reviews*, *41*(1), 1.

Maddern, S. W. (2013, 4 February). Melting pot theory. In *The Encyclopedia of Global Human Migration*. Retrieved 1 February 2022 from https://doi.org/ https://doi.org/10.1002/9781444351071.wbeghm359

Marx, K., & Engels, F. (1885). *Das Kapital*. Verlag von Otto Meissner.

Moreno, C. (2017). Beyond the smart city. PCA-Stream. Retrieved 5 December 2020 from www.pca-stream.com/en/articles/carlos-moreno-beyond-the-smart -city-114

MSCI. (2020). Real estate market size 2019/2020 report. MSCI. Retrieved 1 February 2022 from www.msci.com/documents/1296102/19878845/MSCI _Real_Estate_Market_Size_2020.pdf/06a13e2c-0230-f253-26fa-3318cecb1c59 #:~:text=size%20weights-,The%20size%20of%20the%20professionally %20managed%20global%20real%20estate%20investment,the%20market %20grew%20by%204.1%25

Mycoo, M., & Donovan, M. (2017). A blue urban agenda: Adapting to climate change in the coastal cities of Caribbean and Pacific small island developing states. Retrieved 1 February 2022 from https://publications.iadb.org/publications/ english/document/A-Blue-Urban-Agenda--Adapting-to-Climate-Change-in-the -Coastal-Cities-of-Caribbean-and-Pacific-Small-Island-Developing-States.pdf

Nijman, J., & Wei, Y. D. (2020). Urban inequalities in the 21st century economy. *Applied Geography*, *117*, 102188.

Nishio, A. (2021, 10 August). Facing substantial investment needs, develop- ing countries must sustainably manage debt. Retrieved 1 February 2022 from https://blogs.worldbank.org/voices/facing-substantial-investment-needs -developing-countries-must-sustainably-manage-debt

OECD. (2014). Cities and climate change: National governments enabling local action. In *OECD Policy Perspectives* (pp. 1–21). OECD Publishing.

Oxfam International. (2022, 17 January). Ten richest men double their fortunes in pandemic while incomes of 99 percent of humanity fall. Oxfam International. Retrieved 1 February 2022 from www.oxfam.org/en/press-releases/ten-richest -men-double-their-fortunes-pandemic-while-incomes-99-percent-humanity

Pandey, N. (2016, 16 January). Smart cities could result in social inequality, say experts. Hindu Business Line. Retrieved 11 November 2020 from www.thehindubusinessline.com/economy/smart-cities-could-result-in-social -inequality-say-experts/article9111629.ece

Peterson-Withorn, C. (2021, 30 April). How much money America's billionaires have made during the COVID-19 pandemic. Forbes. Retrieved 1 February 2022 from www.forbes.com/sites/chasewithorn/2021/04/30/american-billionaires -have-gotten-12-trillion-richer-during-the-pandemic/?sh=13af41af557e

Philippe, D., & Haigh, M. (2021, 30 November). Why the world needs a fresh take on smart and sustainable infrastructure. World Economic Forum. Retrieved 2 February 2021 from www.weforum.org/agenda/2021/11/smart -sustainable-infrastructure/

Restrepo, J. D. C., & Morales-Pinzón, T. (2018). Urban metabolism and sustainability: Precedents, genesis and research perspectives. *Resources, Conservation and Recycling, 131*, 216–224.

Ribeiro, H. V., Rybski, D., & Kropp, J. P. (2019). Effects of changing population or density on urban carbon dioxide emissions. *Nature Communications, 10*(1), 3204.

Rodríguez-Pose, A., & von Berlepsch, V. (2019). Does population diversity matter for economic development in the very long term? Historic migration, diversity and county wealth in the US. *European Journal of Population, 35*(5), 873–911.

Schlanger, Z. (2018, 21 February). Every one of Europe's 571 cities is destined for worse heat waves, droughts, or floods. Quartz. Retrieved 4 February 2021 from https://qz.com/1212443/climate-changes-impact-on-europe-all-571 -cities-are-destined-for-worse-heat-waves-droughts-or-floods/

Shaaban, K. (2020). Why don't people ride bicycles in high-income developing countries, and can bike-sharing be the solution? The case of Qatar. *Sustainability, 12*(4). https://doi.org/10.3390/su12041693

Sharifi, A., & Allam, Z. (2021). On the taxonomy of smart city indicators and their alignment with sustainability and resilience. *Environment and Planning B: Urban Analytics and City Science*, 23998083211058798.

Sharifi, A., & Khavarian-Garmsir, A. R. (2020). The COVID-19 pandemic: Impacts on cities and major lessons for urban planning, design, and management. *Science of the Total Environment, 749*, 142391.

Shutters, S., Applegate, J., Wentz, E., & Batty, M. (2021). Scaling of inequality: Urbanization favors high wage earners. *SSRN Electronic Journal*. https://doi .org/10.2139/ssrn.3778929

Siemens. (n.d.). What is urban sustainability? Siemens. Retrieved 2 February 2022 from https://assets.new.siemens.com/siemens/assets/public.1560756617 .90627521-4620-4b1d-9dc6-d94563b93a46.what-is-urban-sustainability-v1.pdf

Silva, B. N., Khan, M., & Han, K. (2018). Towards sustainable smart cities: A review of trends, architectures, components, and open challenges in smart cities. *Sustainable Cities and Society, 38*, 697–713.

Sitinjak, E., Meidityawati, B., Ichwan, R., Onggosandojo, N., & Aryani, P. (2018). Enhancing urban resilience through technology and social media: Case study of urban Jakarta. *Procedia Engineering, 212*, 222–229.

Stamatiadis, N., Pappalardo, G., & Cafiso, S. (2017, 26–28 June). Use of technology to improve bicycle mobility in smart cities. 2017 5th IEEE International Conference on Models and Technologies for Intelligent Transportation Systems, Napoli.

Sumantran, V., Fine, C., & Gonsalvez, D. (2017, 16 March). Our cities need fewer cars, not cleaner cars. *Guardian*. Retrieved 15 June 2021 from www.theguardian.com/environment/2017/oct/16/our-cities-need-fewer-cars-not-cleaner-cars-electric-green-transport

Tan, S. Y., & Taeihagh, A. (2020). Smart city governance in developing countries: A systematic literature review. *Sustainability, 12*(3). https://doi.org/10.3390/su12030899

Thaler, T., Witte, P. A., Hartmann, T., & Geertman, S. C. M. (2021). Smart urban governance for climate change adaptation. *Urban Planning*, 6(3). www.cogitatiopress.com/urbanplanning/article/view/4613/4613

THALES. (2021, 22 April). It's time for a new approach to IoT power consumption. THALES. Retrieved 2 February 2022 from www.thalesgroup.com/en/worldwide-digital-identity-and-security/iot/magazine/its-time-new-approach-iot-power-consumption

UN Chronicle. (2021). Women … in the shadow of climate change. United Nations. Retrieved 26 August 2021 from www.un.org/en/chronicle/article/womenin-shadow-climate-change

UN-Habitat. (2020). World Cities Report 2020: The Value of Sustainable Urbanization. UN-Habitats. Retrieved from https://unhabitat.org/sites/default/files/2020/10/wcr_2020_report.pdf

UNCTAD. (2021). Developing country external debt: From growing sustainability concerns to potential crisis in the time of COVID-19. UNCTAD. Retrieved 2 November 2021 from https://sdgpulse.unctad.org/debt-sustainability/

UNEP. (2014). Emerging issues for small island developing states. UNEP. Retrieved from https://sustainabledevelopment.un.org/content/documents/1693UNEP.pdf

UNEP. (2022). Resource efficiency and green economy. United Nations Environment Programme. Retrieved 1 February 2022 from www.unep.org/explore-topics/resource-efficiency/what-we-do/cities/resource-efficiency-green-economy#:~:text=Cities%20occupy%20only%202%20per,80%20percent%20projected%20for%202050

UNFCCC. (n.d.). Vulnerability and adaptation to climate change in small island developing states. UNFCCC. Retrieved 8 August 2021 from https://unfccc.int/files/adaptation/adverse_effects_and_response_measures_art_48/application/pdf/200702_sids_adaptation_bg.pdf

UNFCCC. (2015). The Paris Agreement. UNFCCC. Retrieved 2 February 2022 from https://unfccc.int/process-and-meetings/the-paris-agreement/the

-paris-agreement#:~:text=The%20Paris%20Agreement%20is%20a,compared
%20to%20pre%2Dindustrial%20levels

UNFCCC. (2021a, 26 February). 'Climate commitments not on track to meet Paris
Agreement goals' as NDC synthesis report is published. UNFCCC. Retrieved
24 September 2021 from https://unfccc.int/news/climate-commitments-not
-on-track-to-meet-paris-agreement-goals-as-ndc-synthesis-report-is-published

UNFCCC. (2021b). Glasgow Climate Change Conference. UNFCCC. Retrieved
5 November 2021 from https://unfccc.int/conference/glasgow-climate-change
-conference-october-november-2021

United Nations. (2011). Transforming Our World: The 2030 Agenda for
Sustainable Development (A/RES/70/1). United Nations. Retrieved 1 February
2022 from https://sustainabledevelopment.un.org/content/documents/
21252030%20Agenda%20for%20Sustainable%20Development%20web.pdf

United Nations. (2015). Goal 5: Achieve gender equality and empower all
women and girls. United Nations. Retrieved 5 February 2022 from www.un
.org/sustainabledevelopment/gender-equality/

United Nations. (2017). New Urban Agenda. H. I. Secretariat. Retrieved 1
February 2022 from https://uploads.habitat3.org/hb3/NUA-English.pdf

United Nations. (2021). Vaccinations and COVID-19 funding for small island
developing states. United Nations. Retrieved 7 August 2021 from www.un
.org/ohrlls/content/covid-19-sids

United Nations Department of Economic and Social Affairs. (2021a). Goal
5: Achieve gender equality and empower all women and girls. UNDESA.
Retrieved 5 February 2022 from https://sdgs.un.org/goals/goal5

United Nations Department of Economic and Social Affairs. (2021b). Goal 11:
Make cities and human settlements inclusive, safe, resilient and sustainable.
UNDESA. Retrieved 4 February 2022 from https://sdgs.un.org/goals/goal11

United Nations Development Programme. (2015). Sustainable Development
Goals. UNDP. Retrieved 1 February 2022 from www.undp.org/content/dam/
undp/library/corporate/brochure/SDGs_Booklet_Web_En.pdf

Valcárcel-Aguiar, B., Murias, P., & Rodríguez-González, D. (2019). Sustainable
urban liveability: A practical proposal based on a composite indicator.
Sustainability, 11(1). https://doi.org/10.3390/su11010086

Wang, S., & Ge, M. (2019, 16 October). About the fastest-growing source of
global emissions: transport. WRI. Retrieved 1 February 2022 from www
.wri.org/insights/everything-you-need-know-about-fastest-growing-source-global
-emissions-transport#:~:text=Emissions%20from%20the%20transport%20sector
,emissions%20from%20burning%20fossil%20fuels

World Bank. (2020a, 7 October). COVID-19 to add as many as 150 million
extreme poor by 2021. World Bank. Retrieved 1 February 2022 from www
.worldbank.org/en/news/press-release/2020/10/07/covid-19-to-add-as-many-as
-150-million-extreme-poor-by-2021

World Bank. (2020b, 8 June). The global economic outlook during the COVID-19
pandemic: A changed world. World Bank. Retrieved 22 April 2021 from
www.worldbank.org/en/news/feature/2020/06/08/the-global-economic-outlook
-during-the-covid-19-pandemic-a-changed-world

World Bank. (2021a). Global Economic Prospects. Retrieved 1 February 2022 from https://openknowledge.worldbank.org/bitstream/handle/10986/35647/9781464816659.pdf

World Bank. (2021b). What a waste 2.0: A global snapshot of solid waste management to 2050. World Bank. Retrieved 1 February 2022 from https://datatopics.worldbank.org/what-a-waste/trends_in_solid_waste_management.html

World History Encyclopedia. (2021). *City Timeline*. World History Encyclopedia. Retrieved 27 January 2022 from www.worldhistory.org/timeline/city/

Zaheer, A. (2020). Sustainability and resilience in megacities through energy diversification, land fragmentation and fiscal mechanisms. *Sustainable Cities and Society*, *53*, 101841.

Zarrilli, S., & Aydiner-Avsar, N. (2020, 13 May). COVID-19 puts women working in SIDS tourism industry at risk. UNCTAD. Retrieved 16 September 2021 from https://unctad.org/fr/node/2415

4. Smart cities as an urban regeneration avenue: redefining the efficiency and performance of cities

INTRODUCTION

The global population has been growing steadily over the past few centuries, particularly in the twentieth and twenty-first centuries. In the twentieth century, the number of people increased from 1.6 billion to 6 billion (Ehrlich, 1995). Such growth was stimulated by factors like technology advancement in different fields such as health, agriculture, education, building and construction to name a few (Council et al., 2000). In the health sector, technology allowed for endemic problems such as high mortality rates, tropical diseases and others to be mitigated (Allam et al., 2019, 2020). In the agricultural sector, technologies such as irrigation, mechanisation and use of chemicals and fertilisers became pronounced, allowing for large-scale and more efficient farming (Paarlberg & Paarlberg, 2008). This prompted an increase in food supply, not only in farming fields, but also in markets (mostly in urban areas) as different modes of transportation were emerging and becoming more efficient and less expensive. In the twenty-first century, technology advancement became even more pronounced and robustly deployed in different dimensions, thus promoting more population growth. For instance, between 2000 and 2021, the global population increased by a further 1.9 billion people to reach a high of 7.9 billion (Cleland, 1996). Going forward, it is estimated that by the end of the century, though the growth rate will be significantly slower, the global population will reach a high of 10.9 billion people (Roser, 2019).

The most interesting phenomenon attributed to the population growth, as indicated in a report by the United Nations Human Settlements Programme (UN-Habitat, 2020), is that most people have been more attracted to urban areas in the past two centuries than any time previously. For instance, it is estimated that by the beginning of the twentieth century, only 30 per cent of the global population was living in cities (Roser, 2019). By 2007, the number of urban dwellers had grown to approximately 50 per cent of the global population. By the end of 2021, the urban population was estimated to be in excess of 55 per cent, and the number will further reach 68 per cent by 2050 (UN DESA, 2018). Such increase in urban population stimulated a number of issues including exerting unprecedented pressure on existing cities as available spaces, infrastructure, urban resources and other elements became scarce relative to human density. Another issue that arose is the rapid emergence and growth of new urban areas, prompted by urban sprawl, that prompted the conversion of arable land, forest and other land reserves into human habitable areas (Allam, 2019a, 2020d). The consequences of these two phenomena (population increase and rapid urbanisation) are many and have been widely documented. Among them is the need for urban regeneration to help cities manage to comfortably and efficiently accommodate the increasing population by providing opportunities such as employment, sufficient housing and increased and efficient service delivery in areas like transport, education, administration and health.

Urban regeneration programmes across different cities have over time prompted numerous positives as well as negatives that still need to be addressed. In terms of the positive impacts, one is the economic contribution of cities to global gross domestic product (GDP). Cities are responsible for approximately 80 per cent of the global economy (World Bank, 2020c), and with the foregoing uptake of different technologies, it is expected that such contributions will continue to increase and become mainstream. This could be possible as technologies are allowing for novel ways of conserving and preserving culture and heritage and new and innovative ways of exploiting creative arts. Technology is seen to help integrate the historical urban landscape (associated with both tangible and intangible cultural assets) to proposed urban projects aimed at reviving part of urban 'decaying' fabrics, especially by allowing such 'assets' to become part of the economic mainstay of cities in areas like tourism attraction, job creation and creativity and preservation, etc.

Urban regeneration is further viewed to have prompted a renewed social interest, for example Sustainable Development Goal (SDG) 11. The clarion call on SDG 11 is the need to make cities more inclusive, safe, resilient and sustainable (United Nations Department of Economic and Social Affairs, 2021). Cities such as Singapore, Barcelona and

Hangzhou that have incorporated different aspects of technology in their urban regeneration agendas are reported to have improved their liveability aspects significantly (Argyriou, 2019; Calzada, 2018; Gascó-Hernandez, 2018; Jiang et al., 2020). For instance, in Singapore, urban regeneration programmes in the housing sector, which began in 1968, allowed the government to solve the housing problem significantly – with almost 95 per cent of Singaporeans securing affordable housing (Allam, 2020b; Amirtahmasebi et al., n.d.).

Urban regeneration, however, is in most cases mired by unprecedented negative impacts that if not properly managed may escalate to increase social inequality as well as prompt unsustainable practices (Speck, 2018). One such negative is gentrification, which has been the outcome of most urban renewal programmes in many cities. For instance, in central London, an urban regeneration programme involving a face lift of the River Thames waterfront caused massive gentrification, where most original social groups were replaced by wealthy households (Atkinson, 2000). In the United States, cities such as New York, Seattle and Boston have experienced massive gentrification, mostly due to regeneration in the housing sector prompting huge increases in real estate prices (Artusio et al., 2017). As a result, a significant number of families have been rendered homeless due to an increase in evictions. Besides gentrification, regeneration programmes are argued to be relatively expensive both in terms of finance and resource needs and most cities across the globe might not have sufficient capacity to undertake comprehensive revitalisation programmes (Gibbons et al., 2021). However, the excessive pressure prompted by an increasing urban population and increasing economic needs impels cities to undertake regeneration projects, hence attracting huge financial burdens. Such financial burdens in turn prompt constraints that hamper effective service delivery, causing delays in the completion of investment projects, widening socioeconomic inequality and exposing the public to challenges like traffic congestion, climate change, food insecurity and others. In light of this background, this chapter explores in detail how smart city technology can be incorporated in urban regeneration agendas, especially to fast-track and enhance urban efficiency and performance.

CITIES URGENTLY NEED REGENERATION

The notable contributions of cities in socioeconomic development and growth need to be safeguarded and encouraged to ensure that the current and anticipated urban population is efficiently catered for. However, there is already a full tray of notable challenges that have significant capacity to slow down meaningful contributions and in worst-case scenarios reverse gains already made. Challenges like the impacts of climate change have been identified to pose serious threats to the economy, the social fabric, urban infrastructure and, in some cases, in coastal and low-lying regions, Combined, those change the landscapes of cities (Rezai et al., 2018; UNESCO, 2019; United Nations Department of Economic and Social Affairs, 2019; Wright et al., 2016). The COVID-19 pandemic is another challenge that engulfed cities, and though the outbreak was experienced only recently (in the wake of 2020) (Allam, 2020a, 2020c), the impacts on different urban fronts outpaced most previous pandemics and the long-term impacts are expected to be enormous (UNCTAD, 2020; World Bank, 2020a, 2020c, 2021). For instance, countries across the globe are already struggling to reinstate their economies to the growth path they experienced before the emergence of the pandemic (OECD, 2020; Wijffelaars, 2020). By 2019, despite cases of climate change impacts being reported in different parts of the globe, global economic growth was projected to continue, albeit at a slow but gradual rate. For example, it was at 2.9 per cent in 2019 and was projected to maintain a positive trajectory, reaching a target of 3.3 per cent by 2020 and a further 3.4 per cent by 2021 (International Monetary Fund, 2020). In the same vein, the urban growth rate also maintained a positive trend in 2019 – maintaining a 1.97 per cent growth compared to the 1.68 per cent reported in rural areas in the same period (Federal Reserve Bank of St Louis, 2019). However, when the COVID-19 pandemic occurred, the global and urban GDP plummeted significantly by approximately −4.9 per cent to below the 3.2 per cent pre-pandemic projection (Allam, 2020a). Although the global economy has been reported to grow by approximately 5.6 per cent (OECD, 2021), the trend is just a 'catch-up' to where the economy was before the pandemic.

When these challenges are coupled with others that are often associated with cities, such as inadequate housing, traffic congestion, an increasing number of unemployed people and increasing poverty, they prompt an urgent need for urban regeneration. The objectives, goals and the need for urban regeneration are also well captured in the SDGs, especially Goal 11 (United Nations Department of Economic and Social Affairs, 2021), and the New Urban Agenda (United Nations, 2017), where the main target is to have the human dimensions captured. Previous regeneration programmes focused on revitalising the dwindling economic activities and attractiveness of cities. The main focus was to attract new investment, economic opportunities and more people, especially in diverse cities facing the issue of population decline. However, nowadays, when most cities are reported to host more than 500,000 people (United Nations, 2016), the challenges to be addressed by regeneration are not primarily about retaining or preventing population decline but are geared toward attracting new socioeconomic opportunities. This is important as unemployment and economic opportunities are becoming inversely proportional to the youthful population growth (who are well educated and ready for the job market). For instance, in 2019, before the emergence of the COVID-19 pandemic, the global average of unemployment was approximately 5.36 per cent (equivalent of 267 million people), but it sharply rose to 6.6 per cent by 2020 (World Bank, 2020b). Unsurprisingly, most of those unemployed reside in urban areas, thus making it difficult for cities to eliminate the challenges of urban poverty, slums and informal settlements to address homelessness (Antipova & Momeni, 2021).

Countries with the lowest unemployment rates, lowest housing challenges, efficient transport challenges and higher liveability index are those that have undergone successful regeneration programmes. For instance, the city of Auckland, which was ranked highest in terms of liveability in 2021 (gathering 96 points; Choudhury, 2021), has been at the forefront of implementing diverse regeneration programmes. A case in point is the $133 million project already set aside to rehabilitate the midtown areas of the city (Our Auckland, 2021). The planned regeneration programme is bound to affect urban components such as the town hall, an art gallery, two universities, city parks, streets and others – all catering for varied human dimensions and population segments. The second-ranked city in terms of liveability is Osaka, with 94.6 points (Choudhury, 2021; Economic Intelligence Unit, 2021) and literature covering its regeneration programmes point toward massive investments channelled toward this course. Of those cities and others ranked highest in liveability status, the common denominator is the focus on human-scale regeneration, where the focus is not only on economic activities, but also on areas like education, green spaces and walkable streets.

After the outbreak and unprecedented spread of COVID-19 in 2020 prompted worldwide lockdown and diverse containment measures, it became apparent that most cities, despite being rated in terms of economic growth based on GDP metrics, had fallen short in human dimensions (Sharifi & Khavarian-Garmsir, 2020). This could be factors like food supply shortages, massive job losses – especially for women and youths who are more likely to be working in informal sectors (Antipova & Momeni, 2021; Pak et al., 2020; Zarrilli & Aydiner-Avsar, 2020) – and the untimely closure of learning institutions (Osman, 2020), among others. While the need for lockdowns were justified by the rapid spread and increasing fatalities from the pandemic, it exposed the reality that cities were fashioned in ways that social orders could be disrupted by a wide range of externalities (Allam & Jones, 2020). Adopting urgent regeneration projects in different socioeconomic sectors, however, has the capacity to help expand economic networks, which in turn would allow for the accommodation of people both in formal and informal sectors. Regenerating the cultural heritage sector could have a significant impact on cities, for example the creation of new job opportunities, enhanced innovation in creative arts, the promotion of social inclusion and the promotion of sustainability practices. It has further been argued that urban areas have the potential to find their own identity if they adopt cultural-led urban regeneration. By so doing, cities would experience benefits such as the restoration of social pride, where the building of a unique identity becomes inevitable, hence leading to more close-knit communities (Hwang, 2014). A report by UNESCO highlights that approximately 29.5 million people are employed in the cultural and creative industry globally and the potential in this industry is unlimited if properly tapped (UNESCO, n.d.). While this industry is reported to have been among the first to experience the negative impacts of COVID-19 as most of its sectors were closed down (UNESCO, 2021), it could benefit from the integration of smart technologies such as augmented reality, artificial intelligence and others to make it more resilient. In fact, from the highlights of the contribution of the cultural and creative industry to the global economy (e.g. in 2019 the sector earned the United Kingdom approximately 115.9 billion euros), especially before the pandemic (Waitzman, 2021), it can be among the potential industries that could allow economies, and especially cities, to experience real growth as they explore new frontiers to bolster economic recovery.

From the above discussion, it is evident that urban regeneration needs to be hastened in most cities to not only allow for accommodating an increasing population, but also to help cities transition to sustainable practices such as the use of alternative energies, the adoption of alternative mobility and opening new economic frontiers.

URBAN REGENERATION ANCHORED ON SMART TECHNOLOGIES

In the introduction section of this chapter, it was highlighted that urban regeneration is usually costly and has the potential to attract unpleasant negative outcomes that could be counterproductive for cities. However, the downsides of regeneration could be overcome by embracing and deploying modern technologies that have been tried and tested in other urban planning approaches (Allam, 2019b; Allam et al., 2018). For instance, in the building and construction industry, modern technologies such as robotics, 3D printing, artificial intelligence and others have been deployed to not only increase efficiency but also to lower costs such as labour and raw materials (Deloitte, 2021; Global Infrastructure Hub, 2021; Jefferson Online, 2017; Rahman, 2020). In the transport sector, in the pursuit of identifying new and alternative mobility options, technologies such as big data, the Internet of Things, artificial intelligence and crowd computing are widely deployed. The targets are to enhance security, reduce traffic incidences, promote safety for those using bicycles and allow for ride sharing, etc. (Biyik et al., 2021; Dixit et al., 2021). Likewise, modern technologies are known to have the potential to solve a wide range of challenges associated with urban regeneration. For instance, in the previous section it was noted how technologies such as augmented reality, digital twins and artificial intelligence could be deployed in the cultural and creative industry to help achieve a number of objectives. These include conservation of both tangible and intangible cultural assets (such as artefacts) and built infrastructures (such as buildings, monuments and others). The use of such technologies has been the subject of research in many different areas and positive results have been obtained. For instance, in Brasov, Romania, Voinea et al. (2019) adopted augmented reality technology coupled with Google's Tango and ARCore platforms to show how it could be used in the conservation and promotion of cultural heritage in old buildings. Țara Bârsei in Brasov, a UNESCO-listed monument, was used as a case study, where a 3D representation of the physical object was constructed. The outcomes of the study showed that the different technologies, especially augmented reality, were increasing users' experiences and enjoyment of the cultural assets and had unmatched potential in transforming cultural heritage conservation efforts, as well as providing access – via digital means – to a broader population.

Besides the physical application of technologies in actual regeneration programmes, it has been argued that modern digital tools have the potential to help urban revitalisation in regard to processes and decision making. On this, modern technologies, especially Internet of Things, big data, crowd computing and artificial intelligence, have been deployed in many urban realms to help in the collection of data that are subsequently used to arrive at different decisions (Duan et al., 2019; McGovern et al., 2017; Pencheva et al., 2018; Sarker et al., 2018). In regeneration programmes, data are critical as they allow for the reliable and real-time study of different urban dynamics which could allow for informed decisions in frontiers like housing upgrades, infrastructure expansions, land use, environmental sustainability agendas and others (Kamrowska-ZbBuska & Obracht-ProndzyDska, 2018). Data further play a significant role in helping address different social issues, especially by incorporating the insights and inputs derived from public participation. Involving communities and urban residents in decision making on issues related to cities has been identified as a critical positive step that helps enhance the relationship between decision makers and the public. It allows for the prioritisation of projects and addresses the interests of communities living in areas proposed for regeneration projects. This approach has the potential to help avert possible conflicts, apprehension and rejection of projects. In the modern urban planning dispensation, where smart technologies abound, it is possible and efficient to collect relevant data from the public using different digital tools (Linders, 2012; Yeung, 2018). Those tools were well pronounced during the height of COVID-19 where they seamlessly replaced the traditional participatory approaches that were rendered ineffective by the healthy protocols instituted in different parts of the globe (Landi et al., 2022; Padeiro et al., 2021; Yang et al., 2020). But even before the outbreak of the pandemic, the adoption of digital participatory tools had proven effective in attempts to reach a wide range of stakeholders in a cost-effective manner and in real time. These further allow for seamless collaboration in designing, planning and implementing proposed programmes (Lin, 2022).

Anchoring urban regeneration on smart technologies is further argued to help planners map and incorporate current and future trends in their revitalisation agendas. A case in point is the rise and widespread use of digital twin technology. Previously, this technology was highly appraised in the manufacturing sector as it allowed for the virtual representation of different components, hence allowing for remote access and interaction with the component in question as in the physical realm (Deren et al.,

2021). Similar approaches of representing the physical object (in this case the urban area that is earmarked for regeneration) in the virtual environment has been deemed possible in urban planning. Proponents of this technology are particularly optimistic for its potential as a 'testbed' for different scenarios before the actual regeneration work is undertaken in physical locations (Allam & Jones, 2021). The technology allows for unlimited simulations in terms of design, feasibility and delivery approaches that are critical in estimating not only investment costs but also the estimation of social and economic dividends that could be derived from the projects (Klebanov et al., 2018; Yeh & Li, 2002; Zhang et al., 2021).

Another reason for anchoring urban regeneration to smart technologies is the fact that most of the revitalisation needs are prompted by the dilapidation of certain urban components such as infrastructure, buildings and services as a result of the emergence of new technologies. Therefore, it becomes logical to align regeneration to those technologies. For instance, with regard to communication and social interactions, most prominent options are the use of social media and virtual communication tools that rely on high-speed communication networks. In such cases, the adoption of technologies such as 4G, 5G and other future mobile telecommunication networks is inevitable (Allam & Jones, 2021). While these are adopted to fill the need for communication, they also help with regeneration, especially through their influence on data collection and sharing in different urban frontiers.

STAKEHOLDER PARTICIPATION IN SMART REGENERATION

In the course of implementing urban regeneration programmes, cities across the globe have had to put up with a wide range of challenges, ranging from financial problems, scarcity of land and legal and ethical challenges, especially when the programs directly impact social aspects like housing (Fenster & Kulka, 2016; Schubert, 2019). However, the most critical challenge is getting every stakeholder to agree with the proposed revitalisation decision. This challenge is noted to undermine even novel programmes that have huge socioeconomic dividends for urban residents, and in most cases it is tied to misunderstandings during the planning stages (Arnstein, 1969; Layson & Nankai, 2015). On this, it has been proposed that the most optimal remedy is ensuring that there is sufficient public participation that captures the aspirations, interests and specific needs of the different participants (stakeholders) (Irvin & Stansbury, 2004; Pereira et al., 2017). The controversy arises in defining who the main stakeholders are, as in most cases the general public, especially minority groups and the poor, do not always get the opportunity to voice their concerns about new urban developments in their locality (Arnstein, 1969). This argument is affirmed by increasing incidences of social exclusion in terms of economic opportunities, where the economic inequality gap is seen to be widening even as urban areas undergo massive regeneration exercises (Beall, 2002; McGranahan et al., 2016). This exclusion is further exacerbated by the increasing cases of homelessness, increasing number of urban poor and an insufficient supply of basic resources and services in different cities across the globe. Some of these challenges could be solved by actively engaging the 'victims' and allowing them to propose some of the solutions to address their conditions. Such solutions could be in the form of novel land use especially in residential areas to facilitate active housing programmes that would cater for all demographic categories. The proposed solutions could further help address challenges like gentrification that often renders minority groups, the poor and low-income earners homeless, or exposes them to economic challenges like a high cost of living.

One of the main objectives of SDG 11 is to make cities and human settlements inclusive so as to achieve safety, resilience and sustainability. In order to facilitate public participation, however, there is a need for a paradigm shift in the decision-making approach. In most cases, the top-down approach has been favoured by those in leadership positions, for their own interests (Semeraro et al., 2020). However, this approach has often led to little or no attention to those below them on the ladder. On the other hand, the bottom-up planning approach has been championed as one with the capacity to allow for the accommodation of all stakeholders, by taking into consideration their inputs on priority areas, approaches to be taken and the nature of projects to be implemented (Layson & Nankai, 2015).

With respect to modern urban regeneration programmes where the adoption of smart technologies is seen to gain traction in planning, incorporating and institutionalising the public participation approach is inevitable. This is based on the fact that most of the smart regeneration programmes heavily rely on the availability of data – where the citizens are part of the most important constituent in providing quality and real-time data. The increase in the number of social platforms has allowed massive flow of data from the citizens that increase the capacity to influence public decision making. The data are either accessed directly from structured public participation channels or indirectly from social media platforms. Either way, they have been a source of valuable information. While it is possible for local governments to obtain substantial data from the installed smart 'things' such as sensors, cameras, drones, etc., they would benefit more by collating this with data from different stakeholders, especially the general public. Further, on the issue of data use, one of the main challenges that has been identified and argued to have potential to compromise the quality and reliability of insights is the case of data privacy and security. These challenges arise from a number of factors, including the overreliance on third parties to collect, store, compute and analyse public data (Ismagilova et al., 2020). As a result, local governments have little or no capacity to monitor or control how such data are managed.

The other factor is related to the ethical and moral concerns that may arise from the use of private data without the consent of individuals whose data are being managed by third parties. The aspect of ethics and morality of data use has the potential to derail the implementation and completion of projects and programmes, especially when legal redress is instituted. This in turn has the potential to water down the reputation of the projects or divert financial resources to cater for legal matters, thus making projects more costly and prompting prolonged delays or abandonment. Public participation by varied stakeholders in urban regeneration programmes is important especially in the case where the projects are being undertaken by the private sector under public-private partnership schemes. The bottom line for private-sector participants is the maximisation of profits. On the other hand, the contracting governments and general public want benefits that influence both the people and the planet (Ng et al., 2010). In order to achieve a balanced outcome for all the parties, public participation is championed as it allows all the parties to be in agreement on different issues like financing, expected outcomes, the duration of the project and terms and conditions for compensation for those affected among others (Nederhand & Klijn, 2016). It has also been argued that stakeholder participation in public-private partnership regeneration programmes allow the private sector to present projects that incorporate innovativeness and those that allow for risks, costs and benefits to be mutually shared.

The role of public participation in regeneration projects in cities cannot be substituted or ignored, especially in the current climate where a wide range of platforms could be adopted to achieve them. Even on sensitive matters relating to finance, the need for public participation is paramount even if this means restructuring participation to safeguard the integrity of contracts. After all, most regeneration programmes are undertaken for the sake of the general public and welfare of the environment, as envisioned in the SDGs.

CHALLENGES FACING SMART URBAN REGENERATION

The emergence of the fourth industrial revolution (Industry 4.0) increased the impetus by cities to pursue the integration and deployment of information and communication technologies in their fabrics. In the recent past, especially post-2015 after the launch of global policy documents and accords in respect to sustainability (such as the Paris Agreement), human urban settlement and socioeconomic agendas (like the SDGs, the New Urban Agenda and the 2030 Agenda for Sustainable Development), there has been an increased focus on integrating technologies on a different urban frontier, the intention being to develop and promote sustainability, pursue inclusivity, increase efficiency and performance and promote liveability status. While some of those objectives were part of the urban regeneration agenda even before the emergence of Industry 4.0, different technologies are seen to help fast-track the intended outcomes. For instance, the aspect of renewing the urban housing sector to help cities accommodate the rapidly increasing population is seen to benefit from technologies such as automation and robotics that make pre-fabrication easier and cost effective (Dawson, 2021; Global Infrastructure Hub, 2021; Macrorie et al., 2019). This promotes mass housing projects as feasible solutions in many cities. In the energy sector, where most global attention has been trained following an increase in the amount of emissions, different smart technologies are helping in the adoption of alternative energy production (Bachner et al., 2019; Bungane, 2017; Chel & Kaushik, 2018; IRENA, 2019). For instance, technologies have made it possible to lower the cost of producing solar components and subsequent installation of the same even on small-scale platforms like rooftops, as well as increasing efficiency in energy consumption through the use of smart grids, smart meters, smart housing programmes and others (Alam et al., 2017; Amin & Schewe, 2017; Bhatti & Danilovic, 2018).

Despite the numerous positives that technologies have made possible to the urban regeneration agenda, they have equally faced substantial challenges that have had the potential to derail their success. Such challenges have been widely documented following their impacts and capacities to render regeneration programmes counterproductive. The most notable issues include scarcity and complexity in securing financial resources, cybersecurity risks, unhealthy competition among smart services and product providers and the newness of most technologies (Bertolini et al., 2020; Deloitte, 2018; Ismagilova et al., 2020). With regard to financing, it is important to note that most regeneration programmes are initiated with the aim of reviving the socioeconomic aspects of target areas or an entire city that have experienced massive job outflow, reduced capital investments and outflow of people and businesses (Hwang, 2014; Kim et al., 2020). Such areas demand paradigm shifts with substantial financial and resource requirements. However, only a few urban areas have the capacity to undertake massive development projects relying on public budgets (Deloitte, 2018; Majid & Scheker-Izquierdo, 2019). Therefore, for cities that are determined to initiate regeneration programmes, the most potent route has been debt financing (Deloitte, 2018), which is not only expensive but has the potential to negatively impact service delivery, risk public property and promote negative trends like side-tracking public participation.

Another challenge that smart urban regeneration is facing is related to increasing social and economic exclusion that already exist in most cities. This is paradoxical as smart technologies have been fronted as tools with the potential to assist in reducing inequalities by opening new economic frontiers in different sectors. For instance, in the transport sector, smart technologies have allowed for the emergence of startups offering ride-sharing services, bicycle renting, paperless transactions and many other similar digital innovations (Gambella et al., 2019; Shaaban, 2020). These have in turn helped to open up new employment opportunities, thus promoting both economic inclusion as well as some elements of sustainability. However, these same technologies have the potential to increase exclusion, more so in areas where cases of the digital divide are more pronounced. In research conducted in Beijing, China, involving 1,195 taxi drivers, on how ride-sharing apps influence equality and wellbeing in the sharing economy, it was established that they are influenced by factors like income, access to technologies and peer adoption (Liu & Xu, 2018). In terms of access to technologies, it is reported that approximately 28 per cent of global urban households are yet to be connected to the internet. This is true even in developed countries where approximately 13 per cent are yet to be connected (UN-Habitat, n.d.). Besides internet connectivity, a sizeable number of the global urban population do not have devices such as smartphones that would allow them access to the internet. As of 2020, there were approximately 6.4 billion smartphones, meaning that over 1.4 billion people did not have access to smart services via the phone (O'Dea, 2021). This is significant, noting that approximately 55 per cent of the global population already resides in cities.

On the economic front, it has rightly been argued that urban regeneration often leads to gentrification where the poor and low-income earners are forced out of their original neighbourhoods as they are unable to cope with increasing rates (Atkinson, 2000). The aftermath of such forced migration is an increased number of urban poor, increased challenges on housing, inefficient service delivery, especially for those in informal settlements, and others (Iyanda & Lu, 2021; Speck, 2018). It is also associated with increasing mental health problems in urban areas where those disadvantaged by revitalisation projects are seen to suffer mental challenges (Corcoran, 2020). However, this should not be the case as urban regeneration and renewals have been deemed as among the approaches that could be adopted to prevent and rehabilitate mental health. This means that proper planning needs to be initiated to ensure that the regeneration programmes proposed are executed in an inclusive way such that even those affected are compensated or have new opportunities to improve their socioeconomic wellbeing (McGowan & Qi, 2020).

Another major issue that may arise as a result of smart urban regeneration is the aspect of cybersecurity concerns. In particular, in relation to the smart city concept, concerns regarding privacy and security attract substantial attention (Sharif & Pokharel, 2022). This has prompted the need to adopt multipronged approaches to guarantee residents that their data and urban systems are secure. Among such strategies include the adoption of alternative energy production and distribution to ensure that all components installed in urban areas do not suffer from power shortages, or via technical loopholes, rendering pathways for phishing or cyberattacks. Further, the adoption of technologies such as blockchain increases immutability, data encryption, trust and the removal of third parties in the case of smart contracts (Allam, 2018; Allam & Jones, 2019; Iberdrola, 2021). While these technologies and approaches increase the total cost of projects, they can be seen as necessary as they allow residents to embrace the projects initiated and accrue the benefits thereof. Otherwise, it might be a tall order for local governments and urban managers to convince residents to cooperate in projects that they may perceive have the potential to jeopardise their privacy and security.

REFERENCES

Alam, M., St-Hilaire, M., & Kunz, T. (2017). An optimal P2P energy trading model for smart homes in the smart grid. *Energy Efficiency, 10*(6), 1475–1493.

Allam, Z. (2018). On smart contracts and organisational performance: A review of smart contracts through the blockchain technology. *Review of Economic and Business Studies, 11*(2), 137–156.

Allam, Z. (2019a). The city of the living or the dead: On the ethics and morality of land use for graveyards in a rapidly urbanised world. *Land Use Policy, 87,* 104037.

Allam, Z. (2019b). Identified priorities for smart urban regeneration: Focus group findings from the city of Port Louis, Mauritius. *Journal of Urban Regeneration Renewal, 12*(4), 376–389.

Allam, Z. (2020a). The first 50 days of COVID-19: A detailed chronological timeline and extensive review of literature documenting the pandemic. *Surveying the COVID-19 Pandemic and Its Implications, 1,* 1–7.

Allam, Z. (2020b). The rise of Singapore. In *Urban Governance and Smart City Planning.* Emerald Publishing.

Allam, Z. (2020c). The second 50 days: A detailed chronological timeline and extensive review of literature documenting the COVID-19 pandemic from day 50 to day 100. *Surveying the COVID-19 Pandemic and Its Implications. Elsevier Public Health Emergency Collection, 9,* 9–39.

Allam, Z. (2020d). Urban and graveyard sprawl: The unsustainability of death. In *Theology and Urban Sustainability* (pp. 37–52). Springer.

Allam, Z., & Jones, D. S. (2019). The potential of blockchain within air rights development as a prevention measure against urban sprawl. *Urban Science, 3*(1), 38.

Allam, Z., & Jones, D. S. (2020). Pandemic stricken cities on lockdown: Where are our planning and design professionals (now, then and into the future)? *Land Use Policy, 97,* 104805.

Allam, Z., & Jones, D. S. (2021). Future (post-COVID) digital, smart and sustainable cities in the wake of 6G: Digital twins, immersive realities and new urban economies. *Land Use Policy, 101,* 105201.

Allam, Z., Dhunny, A. Z., Siew, G., & Jones, D. S. J. S. C. (2018). Towards smart urban regeneration: Findings of an urban footprint survey in Port Louis, Mauritius. *Smart Cities, 1*(1), 121–133.

Allam, Z., Tegally, H., & Thondoo, M. (2019). Redefining the use of big data in urban health for increased liveability in smart cities. *Smart Cities, 2*(2), 259–268.

Allam, Z., Dey, G., & Jones, D. S. (2020). Artificial intelligence (AI) provided early detection of the Coronavirus (COVID-19) in China and will influence future urban health policy internationally. *AI, 1*(2), 156–165. www.mdpi.com/ 2673-2688/1/2/9

Amin, M., & Schewe, P. F. (2017, 14 August). Preventing blackouts: Building a smarter power grid. Scientific American. Retrieved 12 March 2019 from www .scientificamerican.com/article/preventing-blackouts-power-grid/?redirect=1

Amirtahmasebi, R., Orloff, M., & Wahba, S. (n.d.). Singapore. World Bank. Retrieved 21 February 2022 from https://urban-regeneration.worldbank.org/node/72#:~:text=The%20Urban%20Renewal%20Unit%20was,heart%20of%20Singapore's%20finance%20sector

Antipova, A., & Momeni, E. (2021). Unemployment in socially disadvantaged communities in Tennessee, US, during the COVID-19. *Frontiers, 3.* https://doi.org/10.3389/frsc.2021.726489

Argyriou, I. (2019). 9 - The smart city of Hangzhou, China: The case of Dream Town Internet village. In L. Anthopoulos (Ed.), *Smart City Emergence* (pp. 195–218). Elsevier.

Arnstein, S. R. (1969). A ladder of citizen participation. *Journal of the American Institute of Planners, 35*(4), 216–224.

Artusio, M., Axelrod, Z., Davis, B., Danielle Dong, Elzey, C., Joye, J., Kim, A. S., Kone, J., Nelson, G. A., Randall, K., & Roman, A. S. (2017). *Comparative Gentrification Policy: Displacement, Housing Instability, and Homelessness.* S. o. D. University of Pennsylvania, Department of City Planning. Retrieved from www.housingconsortium.org/wp-content/uploads/2021/04/GentrificationStudio_WorkingPaper_Final_012618.pdf

Atkinson, R. (2000). The hidden costs of gentrification: Displacement in central London. *Journal of Housing and the Built Environment, 15*(4), 307–326.

Bachner, G., Steininger, K. W., Williges, K., & Tuerk, A. (2019). The economy-wide effects of large-scale renewable electricity expansion in Europe: The role of integration costs. *Renewable Energy, 134,* 1369–1380.

Beall, J. (2002). Globalization and social exclusion in cities: Framing the debate with lessons from Africa and Asia. *Environment and Urbanization, 14*(1), 41–51.

Bertolini, M., Buso, M., & Greco, L. (2020). Competition in smart distribution grids. *Energy Policy, 145,* 111729.

Bhatti, H. J., & Danilovic, M. (2018). Making the world more sustainable: Enabling localized energy generation and distribution on decentralized smart grid systems. *World Journal of Engineering and Technology, 6,* 350–382.

Biyik, C., Allam, Z., Pieri, G., Moroni, D., O'fraifer, M., O'Connell, E., Olariu, S., & Khalid, M. (2021). Smart parking systems: Reviewing the literature, architecture and ways forward. *Smart Cities, 4*(2), 623–642.

Bungane, B. (2017, 8 March). Renewable energy driving mega cities' transformation. ESI Africa. Retrieved 20 January 2019 from www.esi-africa.com/renewable-energy-driving-mega-cities-transformation/

Calzada, I. (2018). (Smart) citizens from data providers to decision-makers? The case study of Barcelona. *Sustainability, 10,* 3252.

Chel, A., & Kaushik, G. (2018). Renewable energy technologies for sustainable development of energy efficient building. *Alexandria Engineering Journal, 57*(2), 655–669.

Choudhury, S. R. (2021, 8 June). These are the world's most liveable cities in 2021. CNBC. Retrieved 22 February 2022 from www.cnbc.com/2021/06/09/global-liveability-index-2021-world-most-liveable-cities.html

Cleland, J. (1996). Population growth in the 21st century: Cause for crisis or celebration? *Tropical Medicine and International Health, 1*(1), 15–26.

Corcoran, R. (2020). Urban regeneration and the mental health and well-being challenge: In support of evidence-based policy. *Journal of Urban Regeneration and Renewal, 13*(3), 257–269.

Council, N. R., Education, C. B. S. S., Population, C., Projections, P. P., Bulatao, R. A., & Bongaarts, J. (2000). *Beyond Six Billion: Forecasting the World's Population*. National Academies Press.

Dawson, L. (2021, 16 March). Construction industry gets automated at new Monash University Facility. Monash University. Retrieved 20 August 2021 from www.monash.edu/news/articles/construction-industry-gets-automated -at-new-monash-university-facility

Deloitte. (2018). The challenge of paying for smart cities projects. Deloitte. Retrieved 23 January 2022 from www2.deloitte.com/content/dam/Deloitte/us/Documents/ public-sector/us-ps-the-challenge-of-paying-for-smart-cities-projects.pdf

Deloitte. (2021). 2021 engineering and construction industry outlook. Deloitte. Retrieved 25 August 2021 from www2.deloitte.com/us/en/pages/energy-and -resources/articles/engineering-and-construction-industry-trends.html

Deren, L., Wenbo, Y., & Zhenfeng, S. (2021). Smart city based on digital twins. *Computational Urban Science, 1*(1), 4.

Dixit, A., Kumar Chidambaram, R., & Allam, Z. (2021). Safety and risk analysis of autonomous vehicles using computer vision and neural networks. *Vehicles, 3*(3). https://doi.org/10.3390/vehicles3030036

Duan, Y., Edwards, J. S., & Dwivedi, Y. K. J. I. J. o. I. M. (2019). Artificial intelligence for decision making in the era of big data: Evolution, challenges and research agenda. *International Journal of Information Management, 48*, 63–71.

Economic Intelligence Unit. (2021). The global liveability index 2021. EIU. Retrieved 22 February 2022 from www.eiu.com/n/campaigns/global -liveability-index-2021/

Ehrlich, P. R. (1995). *The Population Bomb*. Buccaneer Books.

Federal Reserve Bank of St Louis. (2019, 30 July). Looking at the urban–rural divide in economic growth. Federal Reserve Bank St Louis. Retrieved 21 February 2022 from www.stlouisfed.org/on-the-economy/2019/july/looking -urban-rural-divide-economic-growth

Fenster, T., & Kulka, T. (2016). Whose knowledge, whose power? Ethics in urban regeneration projects with communities. *Geografiska Annaler: Series B, Human Geography, 98*(3), 221–238.

Gambella, C., Monteil, J., Dekusar, A., Cabrero Barros, S., Simonetto, A., & Lassoued, Y. (2019). A city-scale IoT-enabled ridesharing platform. *Transportation Letters*, 1–7. https://doi.org/10.1080/19427867.2019.1694206

Gascó-Hernandez, M. (2018). Building a smart city: Lessons from Barcelona. *Communications of the ACM, 61*(4), 50–57.

Gibbons, S., Overman, H., & Sarvimäki, M. (2021). The local economic impacts of regeneration projects: Evidence from UK's single regeneration budget. *Journal of Urban Economics, 122*, 103315.

Global Infrastructure Hub. (2021). Prefabrication to building parts and modular construction. GIH. Retrieved 3 December 2021 from www.gihub.org/

resources/showcase-projects/prefabrication-of-building-parts-and-modular
-construction/

Hwang, K. H. (2014). Finding urban identity through culture-led urban regeneration. *Journal of Urban Management, 3*(1), 67–85.

Iberdrola. (2021). Blockchain technology at the service of urban management. Iberdrola. Retrieved 27 January 2022 from www.iberdrola.com/innovation/blockchain-for-smart-cities-urban-management#:~:text=ADVANTAGES%20OF%20BLOCKCHAIN%20FOR%20GOVERNING%20CITIES&text=Cities%20can%20interconnect%20using%20blockchain,their%20inhabitants%20in%20real%20time

International Monetary Fund. (2020, January). Tentative stabilization, sluggish recovery? International Monetary Fund. Retrieved 23 February 2022 from www.imf.org/en/Publications/WEO/Issues/2020/01/20/weo-update-january2020

IRENA. (2019). Future of solar photovoltaic: Deployment, investment, technology, grid integration and socio-economic aspects. International Renewable Energy Agency. Retrieved 1 February 2022 from https://irena.org/-/media/Files/IRENA/Agency/Publication/2019/Nov/IRENA_Future_of_Solar_PV_2019.pdf

Irvin, R. A., & Stansbury, J. (2004). Citizen participation in decision making: Is it worth the effort? *Public Administration Review, 64*(1), 55–65.

Ismagilova, E., Hughes, L., Rana, N. P., & Dwivedi, Y. K. (2020). Security, privacy and risks within smart cities: Literature review and development of a smart city interaction framework. *Information Systems Frontiers.* https://doi.org/10.1007/s10796-020-10044-1

Iyanda, A. E., & Lu, Y. (2021). 'Gentrification is not improving my health': A mixed-method investigation of chronic health conditions in rapidly changing urban neighborhoods in Austin, Texas. *Journal of Housing and the Built Environment.* https://doi.org/10.1007/s10901-021-09847-8

Jefferson Online. (2017, 16 August). Construction robotics: How it's changing the industry. Thomas Jefferson University. Retrieved 20 August 2021 from https://online.jefferson.edu/construction-management/construction-robotics-changing-industry/

Jiang, H., Geertman, S., & Witte, P. (2020). Smart urban governance: an alternative to technocratic 'smartness'. *GeoJournal.* https://doi.org/10.1007/s10708-020-10326-w

Kamrowska-ZbBuska, D., & Obracht-ProndzyDska, H. J. S. (2018). The use of big data in regenerative planning. *Sustainability, 10*(10), 3668.

Kim, H. W., Aaron McCarty, D., & Lee, J. (2020). Enhancing sustainable urban regeneration through smart technologies: An assessment of local urban regeneration strategic plans in Korea. *Sustainability, 12*(17), 6868.

Klebanov, B., Nemtinov, A., & Zvereva, O. J. I. M. S. G. S. (2018). Simulation as an effective instrument for strategic planning and transformation of smart cities. *18th International Multidisciplinary Scientific GeoConference, 18*(2.1), 685–692.

Landi, S., Costantini, A., Fasan, M., & Bonazzi, M. (2022). Public engagement and dialogic accounting through social media during COVID-19 crisis: A missed opportunity? *Accounting, Auditing and Accountability Journal, 35*(1), 35–47.

Layson, J. P., & Nankai, X. (2015). Public participation and satisfaction in urban regeneration projects in Tanzania: The case of Kariakoo, Dar es Salaam. *Urban, Planning and Transport Research, 3*(1), 68–87.

Lin, Y. (2022). Social media for collaborative planning: A typology of support functions and challenges. *Cities, 125*, 103641.

Linders, D. (2012). From e-government to we-government: Defining a typology for citizen coproduction in the age of social media. *Government Information Quarterly, 29*(4), 446–454.

Liu, X., & Xu, W. (2018). Adoption of ride-sharing apps by Chinese taxi drivers and its implication for the equality and wellbeing in the sharing economy. *Chinese Journal of Communication, 12*(1), 7–24.

Macrorie, R., Marvin, S., & While, A. (2019). Robotics and automation in the city: A research agenda. *Urban Geography, 42*(2), 1–21.

Majid, N., & Scheker-Izquierdo, Y. (2019). Zoning in on urban regeneration. KPMG. Retrieved 1 February 2022 from https://home.kpmg/xx/en/home/insights/2019/07/zoning-in-on-urban-regeneration.html

McGowan, J., & Qi, R. (2020). The English planning system, the built environment and preventative mental healthcare: Identifying gaps in alignment and promoting integration. *Journal of Urban Regeneration and Renewal, 13*(3), 328–336.

McGovern, A., Elmore, K. L., II, D. J. G., Haupt, S. E., Karstens, C. D., Lagerquist, R., Smith, T., & Williams, J. K. (2017). Using artificial intelligence to improve real-time decision-making for high-impact weather. *Bulletin of the American Meteorological Society, 98*(10), 2073–2090.

McGranahan, G., Schensul, D., & Singh, G. (2016). Inclusive urbanization: Can the 2030 Agenda be delivered without it? *Environment and Urbanization, 28*(1), 13–34.

Nederhand, J., & Klijn, E. H. (2016). Stakeholder involvement in public-private partnerships: Its influence on the innovative character of projects and on project performance. *Administration and Society, 51*(8), 1200–1226.

Ng, S., Wong, J., & Wong, K. (2010). *Public Participation in Public Private Partnership Projects – the Way Forward*. WIT Press.

O'Dea, S. (2021, 16 December). Smartphone penetration worldwide 2020. Statista. Retrieved 23 February 2022 from www.statista.com/statistics/203734/global-smartphone-penetration-per-capita-since-2005/

OECD. (2020, 5 June). Building back better: A sustainable, resilient recovery after COVID-19. OECD. Retrieved 24 September 2021 from www.oecd.org/coronavirus/policy-responses/building-back-better-a-sustainable-resilient-recovery-after-covid-19-52b869f5/

OECD. (2021). OECD economic outlook sees recovery continuing but warns of growing imbalances and risks. OECD. Retrieved 25 February 2022 from www.oecd.org/newsroom/oecd-economic-outlook-sees-recovery-continuing-but-warns-of-growing-imbalances-and-risks.htm

Osman, M. E. (2020). Global impact of COVID-19 on education systems: The emergency remote teaching at Sultan Qaboos University. *Journal of Education for Teaching, 46*(4), 463–471.

Our Auckland. (2021, 28 September). Auckland Council lists $133 million cluster of projects to regenerate midtown. Our Auckland. Retrieved 22 February 2022 from https://ourauckland.aucklandcouncil.govt.nz/news/2021/09/auckland -council-lists-133million-cluster-of-projects-to-regenerate-midtown/

Paarlberg, D., & Paarlberg, P. (2008). *The Agricultural Revolution of the 20th Century.* John Wiley & Sons.

Padeiro, M., Bueno-Larraz, B., & Freitas, Â. (2021). Local governments' use of social media during the COVID-19 pandemic: The case of Portugal. *Government Information Quarterly, 38*(4), 101620.

Pak, A., Adegboye, O. A., Adekunle, A. I., Rahman, K. M., McBryde, E. S., & Eisen, D. P. (2020). Economic consequences of the COVID-19 outbreak: The need for epidemic preparedness. *Frontier Public Health, 8*(241).

Pencheva, I., Esteve, M., & Mikhaylov, S. J. (2018). Big data and AI: A transformational shift for government: So, what next for research? *Public Policy and Administration, 35*(1), 24–44.

Pereira, G. V., Cunha, M. A., Lampoltshammer, T. J., Parycek, P., & Testa, M. G. (2017). Increasing collaboration and participation in smart city governance: A cross-case analysis of smart city initiatives. *Journal of Information Technology for Development, 23*(3), 526–553.

Rahman, F. U. (2020). Bricklaying robots in construction. The Constructor. Retrieved 20 August 2021 from https://theconstructor.org/building/bricklaying -robots-construction/552063/

Rezai, A., Taylor, L., & Foley, D. (2018). Economic growth, income distribution, and climate change. *Ecological Economics, 146,* 164–172.

Roser, M. (2019). Future population growth. Our World in Data. Retrieved 21 February 2021 from https://ourworldindata.org/future-population-growth#:~: text=Population%20growth%20by%20world%20region,-More%20than%208 &text=The%20United%20Nations%20projects%20that,at%2010.9%20billion %20by%202100

Sarker, M. N. I., Wu, M., & Hossin, M. A. (2018, 26–28 May). Smart governance through bigdata: Digital transformation of public agencies. 2018 International Conference on Artificial Intelligence and Big Data, Chengdu.

Schubert, D. (2019). Jane Jacobs, cities, urban planning, ethics and value systems. *Cities, 91,* 4–9.

Semeraro, T., Nicola, Z., Lara, A., Sergi Cucinelli, F., & Aretano, R. (2020). A bottom-up and top-down participatory approach to planning and designing local urban development: Evidence from an urban university center. *Land, 9*(4). https://doi.org/10.3390/land9040098

Shaaban, K. (2020). Why don't people ride bicycles in high-income developing countries, and can bike-sharing be the solution? The case of Qatar. *Sustainability, 12*(4). https://doi.org/10.3390/su12041693

Sharif, R. A., & Pokharel, S. (2022). Smart city dimensions and associated risks: Review of literature. *Sustainable Cities and Society, 77,* 103542.

Sharifi, A., & Khavarian-Garmsir, A. R. (2020). The COVID-19 pandemic: Impacts on cities and major lessons for urban planning, design, and management. *Science of the Total Environment, 749,* 142391.

Speck, J. (2018). Fight displacement: Use proven tools to limit the negative impacts of gentrification. In *Walkable City Rules* (pp. 32–33). Island Press.

UN DESA. (2018, 16 May). 68% of the world population projected to live in urban areas by 2050, says UN. United Nations. Retrieved 5 January 2021 from www.un.org/development/desa/en/news/population/2018-revision-of-world-urbanization-prospects.html

UN-Habitat. (2020). World cities report 2020: The value of sustainable urbanization. UN-Habitat. Retrieved 1 February 2022 from https://unhabitat.org/sites/default/files/2020/10/wcr_2020_report.pdf

UN-Habitat. (n.d.). Addressing the digital divide: Taking action towards digital inclusion. UN-Habitat. Retrieved from https://unhabitat.org/sites/default/files/2021/11/addressing_the_digital_divide.pdf

UNCTAD. (2020). The impact of the COVID-19 pandemic on trade and development: Transitioning to a new normal. UNCTAD. Retrieved from https://unctad.org/system/files/official-document/osg2020d1_en.pdf

UNESCO. (2019, 9 August). The ethical principles of climate change. UNESCO. Retrieved 25 September 2021 from https://en.unesco.org/news/ethical-principles-climate-change

UNESCO. (2021). Cultural and creative industries in the face of COVID-19: An economic impact outlook. UNESCO. Retrieved 2 February 2022 from https://en.unesco.org/creativity/publications/cultural-creative-industries-face-covid-19

UNESCO. (n.d.). New report shows cultural and creative industries account for 29.5 million jobs worldwide. UNESCO. Retrieved 22 February 2022 from https://en.unesco.org/news/new-report-shows-cultural-and-creative-industries-account-295-million-jobs-worldwide#:~:text=New%20report%20shows%20cultural%20and%20creative%20industries%20account%20for%2029.5%20million%20jobs%20worldwide,-2%20min

United Nations. (2016). The world's cities in 2016. Retrieved 2 February 2022 from www.un.org/en/development/desa/population/publications/pdf/urbanization/the_worlds_cities_in_2016_data_booklet.pdf

United Nations. (2017). New Urban Agenda. H. I. Secretariat. Retrieved 2 February 2022 from https://uploads.habitat3.org/hb3/NUA-English.pdf

United Nations Department of Economic and Social Affairs. (2019, 27 September). Small island developing states, on the front lines of climate and economic shocks, need greater international assistance. UNDESA. Retrieved 13 August 2021 from www.un.org/development/desa/en/news/sustainable/sids-on-climatechange-front-line-need-more-assistance.html

United Nations Department of Economic and Social Affairs. (2021). Goal 11: Make cities and human settlements inclusive, safe, resilient and sustainable. UNDESA. Retrieved 4 February 2022 from https://sdgs.un.org/goals/goal11

Voinea, G.-D., Girbacia, F., Postelnicu, C. C., & Marto, A. (2019). *Exploring Cultural Heritage Using Augmented Reality through Google's Project Tango and ARCore*. VR Technologies in Cultural Heritage.

Waitzman, F. (2021, 28 October). Impact of government policy on the creative sector. House of Lords Library. Retrieved 22 February 2022 from https://lordslibrary.parliament.uk/impact-of-government-policy-on-the-creative-sector/

Wijffelaars, M. (2020, 6 October). Spain's struggle against COVID-19 hampers its economic recovery. Rabo Bank. Retrieved 5 November 2020 from https://economics.rabobank.com/publications/2020/october/spains-struggle-against-covid-19-hampers-its-economic-recovery/

World Bank. (2020a, 8 June). The global economic outlook during the COVID-19 pandemic: A changed world. World Bank. Retrieved 22 April 2021 from www.worldbank.org/en/news/feature/2020/06/08/the-global-economic-outlook-during-the-covid-19-pandemic-a-changed-world

World Bank. (2020b). Unemployment, total (% of total labor force) modeled ILO estimate. World Bank. Retrieved 5 October 2021 from https://data.worldbank.org/indicator/SL.UEM.TOTL.ZS?end=2019

World Bank. (2020c, 20 April). Urban development: Overview. World Bank. Retrieved 27 September 2021 from www.worldbank.org/en/topic/urbandevelopment/overview#:~:text=With%20more%20than%2080%25%20of,and%20new%20ideas%20to%20emerge

World Bank. (2021, 8 June). COVID-19 to plunge global economy into worst recession since World War II. World Bank. Retrieved 15 April 2021 from www.worldbank.org/en/news/press-release/2020/06/08/covid-19-to-plunge-global-economy-into-worst-recession-since-world-war-ii

Wright, H., Reeves, J., & Huq, S. (2016). Impact of climate change on least developed countries: Are the SDGs possible? *IIED*. https://doi.org/10.13140/RG.2.1.2817.0649

Yang, Y., Deng, W., Zhang, Y., & Mao, Z. (2020). Promoting public engagement during the COVID-19 crisis: How effective is the Wuhan local government's information release? *International Journal of Environmental Research and Public Health, 18*(1), 118.

Yeh, A. G., & Li, X. (2002). Urban simulation using neural networks and cellular automata for land use planning. Symposium on Geospatial Theory, Processing and Applications, Ottawa.

Yeung, D. (2018). Social media as a catalyst for policy action and social change for health and well-being: Viewpoint. *Journal of Medical Internet Research, 20*(3), e94.

Zarrilli, S., & Aydiner-Avsar, N. (2020, 13 May). COVID-19 puts women working in SIDS tourism industry at risk. UNCTAD. Retrieved 16 September 2021 from https://unctad.org/fr/node/2415

Zhang, X., Shen, J., Saini, P. K., Lovati, M., Han, M., Huang, P., & Huang, Z. (2021). Digital twin for accelerating sustainability in positive energy district: A review of simulation tools and applications. *Frontiers, 3*(35). https://doi.org/10.3389/frsc.2021.663269

5. The paradox of safety within data-driven smart cities

INTRODUCTION

The quest for data-driven cities began in the 1970s in Los Angeles after the city's Community Analysis Bureau initiated a programme of using data gathered from computer databases, cluster analyses and infrared aerial photographs to inform policy formulation on a myriad of issues (Brasuell, 2015). Later, in the mid-2000s, the fourth industrial revolution (dubbed Industry 4.0) came into force, making it apparent that cities could be effectively managed by relying on data. The availability of data was gaining traction following an unprecedented increase in the number of Internet of Things (IoT) devices and sensors deployed in cities. The period after 2015, when the majority of the ground-breaking global policies and accords were launched, called for decisive action on areas like sustainability and inclusivity in cities and human settlements (Sustainable Development Goal number 11), and a reduction of the global temperature (Paris Agreement) became critical for data usage. By 2015, the global data collected, stored and consumed were approximately 15.5 zettabytes (See, 2021), but after the different global accords and the subsequent commitment by a majority of the global community, the data started to grow exponentially, increasing to 18, 26, 33, 41 and 64.2 zettabytes in 2016, 2017, 2018, 2019 and 2020, respectively (See, 2021). This growth was catalysed by the extensive adoption of the smart city planning concept in different urban milieus, where technologies were being implemented. Such exponential growth in data stimulated different positive outcomes in cities, including increased liveability status, the emergence and widespread use of alternative mobility options and increased uptake of alternative energy production methods, among other benefits (Allam et al., 2019; Allam, 2020d; Bouton et al., 2017; Kroposki et al., 2018). However, the growth and increased collection, storage and use of data attracted concerns on how private data, especially those bearing personal identifying information, were being utilised by different groups.

The concerns on personal data are legitimate and genuine as a wide range of global communities have become victims of data breaches. For instance, in December 2021, hackers managed to infiltrate the SolarWinds Company based in the United States and maliciously added code that allowed them to spy on the 33,000 customers, companies and organisations that SolarWinds served (Jibilian & Canales, 2021). In another example, in April 2021, there was a data breach on Meta (formerly Facebook), where the personal information of over 533 million users was posted online, exposing sensitive information like phone numbers, locations, names, email addresses and dates of birth (Holmes, 2021). Later in the same year (August 2021) there was another breach on T-Mobile where the critical data of over 54 million previous and existing customers were exposed. Sensitive information like names, social security numbers and drivers' licenses were accessed by the alleged hacker, John Brinns, who confessed to have carried out the breach (Cipriani, 2021).

In smart cities, massive structured and unstructured data are generated by different devices, sensors and cameras installed in diverse urban establishments, products (e.g. automobiles) and infrastructures. Data are also available from wearables, social media platforms and other avenues. Much of the data, including that which contain personal identifying details, are generated without the knowledge or consent of urban dwellers. These, including those collected from residents with their full knowledge and consent, have sometimes been used for unintended objectives including targeted marketing (Zoonen, 2016). As a result, there is notable apprehension on the part of urban residents to openly share their private data, even with local governments for critical projects, fearing that these data might be used for purposes beyond the declared intentions (Granville, 2018).

The privacy and security concerns on data are however a paradox, as it has been argued that data-driven cities have more opportunities to be safe than their counterparts (Allam, 2019a). For instance, in a report by the Economist Intelligence Unit published in 2021 featuring 60 major global cities, it was established that top-ranking cities including Copenhagen, Singapore, Sydney, Toronto and Tokyo ranked high in the use of data (Armstrong, 2021). In the course of ranking those cities, one of the major parameters considered was the aspect of 'smartness', with a focus on cybersecurity, of which all the aforementioned cities are noted to have some levels of readiness for (Economist Intelligence Unit, 2021). Therefore, to qualify and remain true to safe and smart city credentials, as showcased in the report, cybersecurity concerns need to be addressed, or seen to be addressed. However, data security concerns are not only based on cybersecurity but also on the physical security of installed infrastructures and establishments (Depoy et al., 2005). On this, physical attacks against different infrastructures that support data collection, storage and computation, such as internet and communication masts, installations hosting internet servers and networks, might lead to severe consequences. Such consequences include the loss and destruction of data, unreliable and low-quality data (Ghafir et al., 2018) and in other cases increased cost of data which could eventually impact on the success of smart city concepts (Platt, 2012).

In light of this background, this chapter seeks to explore a wide range of issues related to data-driven cities and how these are benefitting from the availability of 'big' data prompted by an increase in diverse smart technologies. Further, the chapter underlines the need for urban managers and other stakeholders to increase their efforts in ensuring that the privacy and security aspects of the data they collect, store and use are guaranteed. This, as will be comprehensively addressed in the succeeding section, is critical as it will impact on how urban residents relate, embrace and participate in proposed smart urban projects, especially in the current climate as world economies are adapting to the 'new normal' prompted by the ravaging impacts of COVID-19. The chapter also delves into the different approaches that urban stakeholders can participate in to ensure that they have a stake in guaranteeing the privacy and security of data. Further, the chapter will consider and discuss in depth some of the possible consequences that might be experienced if such initiatives are not sufficiently pursued.

SECURING THE CITY WITH DATA-DRIVEN APPROACHES

Cities across the globe have evolved and grown at an exponential rate, both in terms of number and populations therein, especially during the twenty-first century. In terms of the number of cities across the globe, different characteristics such as size, form, structure and composition have influenced the urbanisation phenomenon. While different countries use assorted categorisation methods to define what constitutes an urban area, there is now a more universal method proposed by the United

Nations Statistical Commission dubbed the 'degree of urbanisation' (OECD et al., 2021). In summary, this categorisation option proposes that an area qualifies as an urban area when it has a minimum of 50,000 people (>1,500 inhabitants per kilometre square) (European Commission et al., 2020). Using this categorisation, there are over 10,000 cities across the world. With the projected urban population growth (60 per cent of global population by 2030) (UN, 2019), the number of cities will continue to grow exponentially beyond the 10,000 mark. This growth is significant and of interest to policy makers especially in view of both the economic contribution of cities and their contribution to climate change. With regard to economic contribution, cities contribute approximately 80 per cent of the global economy (UNDESA, 2018; World Bank, 2020) and are therefore argued to be the economic engine of global gross domestic product. But, with the current trends in climate change, there are concerns that such contributions might start to plummet, prompting negative effects such as increased poverty, unemployment, homelessness, crime and uncontrollable and unsustainable practices (Bezgrebelna et al., 2021; International Organization for Migration, 2019; Maitre et al., 2018).

In view of this, there have been concerted efforts from different quarters aimed at ensuring that cities maintain their role as economic powerhouses while pursuing sustainable practices. Such objectives are noted in a myriad of global policy documents including the Sustainable Development Goals (especially Goal 11) (UNDESA, 2021), the New Urban Agenda (United Nations, 2017) and the 2030 Agenda for Sustainable Development (United Nations, 2011). Among the efforts being adopted are investing in data-driven smart city planning models, where different aspects of urban planning are based on the availability of data. On this front, cities have openly embraced the use of IoT technology, artificial intelligence, machine learning, crowd computing, big data technologies and others (Allam, 2021; Allam & Dhunny, 2019). This can be attested by the number of different IoT smart devices already on the market. For instance, as of 2015, when most of the global accords and policy documents were launched, there were only 15.41 billion devices installed in different cities globally (Statista Research Department, 2016). However, by the end of 2021, the number of connected devices had increased to more than 35.82 billion, and it is projected that by 2025 the number might increase beyond 75 billion devices (Statista Research Department, 2016). While they have prompted positive outcomes especially in terms of efficiency and performance in cities, there are diverse negative effects associated with tech-enabled devices. Of these, the main concern is safety and privacy of data – for both residents and urban elements (Allam, 2020a).

With respect to the security of urban elements such as physical infrastructures and smart devices, concerns have been raised as to how this would be impacted by negative events such as terrorism, protests, extreme weather conditions, power disruptions and natural disasters (Glaeser & Shapiro, 2001). These concerns demand a collaborative approach between urban planners and other stakeholders to explore and identify different methodologies that would fit to conclusively address the challenges. These include digitally transforming public safety agencies (Cisco, 2016), phasing out antiquated technologies (Koretsky & van Lente, 2020), changing the recruitment strategies of security agencies (Mitchell, 2021) and creating private broadband networks (Partida, 2021). This would further require regular upgrading of existing infrastructure (e.g. that support 4G and allow for 5G) to ensure that they support the fast transfer of data, hence mitigating data loss in case of extreme negative eventualities (Alenezi, 2021).

With regard to the security and privacy of data, especially sensitive personal information, there have been concerns sparked by incidences of hacking, handling of data by third entities and a notable lack of sufficient capacity of many local governments to manage data (Ismagilova et al., 2020; Zarifis et al., 2021). The contracting of private firms and large corporations to manage data for many cities has been a cause of concern for residents fearing monetisation of their data or of being monitored. They also fear the possible misuse of personal identifiable information for purposes beyond managing diverse urban agendas (Allam, 2019b, 2020b; Thenuan & Raina, 2016). The concerns on security and privacy of residents in regard to their data do not necessarily relate to direct threats but instead to innocent activities prompted by an influx of data in cities. It is reported that almost 1.7 megabytes of new data are generated and collected every second by different devices installed in different urban infrastructures and private residences (Bulao, 2022). Similarly, data are also generated by different wearables that are gaining traction, especially those that track health-related issues (Azodo et al., 2020). Of the data generated every second, it is reported that a sizeable number (approximately 41 per cent) is lost or misplaced (Help Net Security, 2020), and this triggers apprehension that many residents have of their privacy or security being compromised. This could be tied to a 2019 incident where the company Huawei was alleged to have had the intention of using its 5G technology to advance espionage agendas (Bond & Kynge, 2018;

Mascitelli & Chung, 2019), resulting in it being restricted in the United States. While there was no conclusive evidence to ascertain the allegation, this provides some background to residents' concerns, especially those who view the use of technology in cities as having the potential to lead to a dystopian future (Zoonen, 2016).

It is now inevitable that data-driven approaches will be used in managing urban affairs. It therefore behoves local governments to be proactive in ensuring that data collected within their jurisdiction are only utilised for specified purposes, that they are secure and are being handled in a way that promotes trust. On this, while it might not be possible to interfere with the running of the private sector contracted to offer data management services, local governments have a mandate to vet firms seeking private opportunities. This would not only help achieve the best outcomes for data use but would further impact on the costs, participation of residents in projects and in promoting local startups as they pursue innovations and solutions for different urban frontiers.

ARE PRIVACY AND PERSONAL DATA SECURE?

Data-driven cities have been hailed for the way diverse issues like climate change, sustainability, housing scarcity, traffic and job creation are handled. It has been widely appreciated that cities are comprised of dynamic complex systems that are prone to both internal and external determinants (Bibri, 2021a). Internal determinants include increasing the urban population (UNDESA, 2018), increasing demand for housing (Bezgrebelna et al., 2021), a need to create job opportunities, traffic challenges (Afrin & Yodo, 2020) and sustainability concerns (Bibri, 2021b; Yigitcanlar et al., 2019) that are synonymous with most cities. On the other hand, external factors that impact cities include climate change, migration (Barrios et al., 2006; Kelman, 2015; Rodríguez-Pose & von Berlepsch, 2019; Yu et al., 2019), economic crises and pandemics. Leveraging on data has become possible to positively influence different decisions on how these factors are handled, thus improving efficiency and performance (Bibri, 2020, 2021a, 2021c). This then warrants the need for robust data collection, storage, computation and analysis to draw informed and concrete decisions. The increased emergence and subsequent adoption of diverse technologies such as artificial intelligence, IoT, big data, cloud computing and machine learning make it possible for cities to generate and manage data. These technologies have had positive impacts on the amount of data being generated in cities. For instance, in

2015, with increased attention on sustainability and inclusivity, it was estimated that a city hosting 1 million people had the capacity to generate approximately 180 million gigabytes of data daily (Cisco, 2015). Following that, the amount of data generated globally started to increase and by 2020 it was estimated that approximately 44 zettabytes of data were generated in cities globally. This is due to increase to a high of 463 exabytes by 2025 as more data-generating devices are installed in cities (Desjardins, 2019).

The significance of and subsequent demand for data has prompted increased competition among different corporations offering data management services and products. The focus of competitors is to gain control of the smart city market, currently valued at approximately $739.78 billion (as of 2020 estimates) and expected to reach a high of $2036.10 billion by 2026 (Mordor Intelligence, 2021). As a result, a number of issues have been identified. First is the issue of unstandardised communication between different devices, riding on the absence of universal frameworks that would help in streamlining communication protocols and standards. As a result, most of the data from different devices are stored in silos rather than in a universal database that could be accessed by all players in the urban planning ecosystems (Allam & Jones, 2020). This leads to the second set of issues appertaining to 'locking out' potential startups (Dawar & Frost, 1999) in the smart city market, which is essentially counterproductive as it hampers effective collaboration in pursuing different digital solutions. Third, the availability of data in silos makes it difficult to ascertain the quality and accuracy of data and this could lead to expensive errors, wrong decisions and attract unwarranted costs in the course of implementing different projects (Strengholt, 2020). This has also led to the popularisation of standalone projects rather than pursuing a 'comprehensive whole' in addressing different urban challenges. The smart city vision does not need to be viewed and applied in isolation (Strinsjö, 2020) but different solutions need to cut across various urban frontiers, as cities are comprised of complex intertwined systems and processes that cannot be easily separated and viewed as independent elements.

The negative issues related to data collection, storage, computation and analysis within data-driven urban planning platforms prompt legitimate concerns on trust, transparency and efficiency. These escalate to concerns on privacy and security of data, especially those with the potential to compromise personal information of residents or expose them to security issues. These concerns are not far-fetched, as evidenced by the presence of some cases of cybersecurity that have already been documented. For instance, in 2015, over 230,000 people in Ukraine were cut off from their electricity grids for a substantial amount of time after the national power grid system was hacked (Zetter, 2016). In 2018, the city of Atlanta was reported to have fallen victim of Ransomware attacks that targeted diverse municipal operations prompting service disruptions for months, resulting in the city spending over $2.6 million to resolve the issues (Newman, 2018). During the height of COVID-19 most people needed to make calls while working from home and as a result virtual communication via platforms such as Zoom became very popular (Allam, 2020c, 2020d, 2020e). However, these became targets for hackers who, in some instances, infiltrated online meetings and conferences with explicit material, hate images and threatening language (Davis, 2020). This added weight to the scepticism of some people regarding urban operations and management shifting to digital platforms.

Despite the vivid threat on privacy and security of data, it is worth noting that there are valuable technologies such as blockchain that could be deployed to mitigate privacy and security concerns (Z. Allam, 2018). One key element to boosting the acceptability of data use in urban planning and thus gain the confidence and participation of residents is assuring them of total security and confidentiality. This is why technologies such as blockchain are being deployed to achieve benefits such as immutability, distributed ledgers, smart contracts and others that enhance trust, transparency, efficiency and performance. In addition to using applications that allow for the anonymisation and encryption of data, stakeholders mandated to manage data on behalf of local governments would further need to synchronise their operations – including ensuring the standardisation of communication protocols and frameworks. This would allow digital solutions to be deployed to address multiple urban challenges rather than each in isolation. With seamless collaboration, it would be possible to overcome the challenges of data silos, restrained access to data, address data quality concerns and other issues that intensify fears on privacy and security. The concerns could further be addressed by ensuring that there is sufficient security on the physical establishments and infrastructures where data are generated, stored, computed and analysed.

BALANCING DATA-CENTRIC APPROACHES FOR SAFER CITIES

Cities across the globe are facing numerous challenges ranging from climate change, health (especially after the outbreak of COVID-19 in 2020), traffic congestion, compromised air quality and the scarcity of basic resources like water and food. Many global cities are also experiencing the unprecedented challenge of a shortage of affordable housing following a rapid increase in the urban population. The population growth has been prompted by factors like migration of people from rural areas in search of opportunities (e.g. health, education, employment and recreation). While cases like COVID-19 emerged recently, most of the aforementioned challenges have been present in cities for decades with no solution yet found to address them. As a result, there were increased efforts offering solutions to most of these urban challenges, but lasting solutions have not yet been found for a majority of the issues. For instance, in the transport sector, a sizeable number of cities have tried numerous alternatives to resolve cases of traffic congestion and emissions, accidents, noise and others associated with the sector without much success. Some alternative options include the introduction of subways, an emphasis on bicycle use and the promotion of public transport, but these have not yet helped reduce the number of private cars in city centres, as is evident in most cities in Europe. For instance, passenger cars increased by approximately 1.8 per cent in 2019 in Europe (European Automobile Manufacturers Association, 2021). In the housing sector there were different trends adopted including the building of high-rise structures (Giyasov & Giyasova, 2018), but these have not allowed cities to resolve cases of homelessness and informal settlements (Bezgrebelna et al., 2021). In fact, according to the latest statistics, it is estimated that there are over 150 million homeless people globally (mostly residing in urban areas), with an additional 1.6 billion people who lack adequate housing (Chamie, 2017). These cases, as noted in Chapter 4, are prompted by issues like gentrification driven by the availability of novel smart technologies. When urban areas are revitalised they attract new wealthy households, which in turn prompts a spike in property prices and cost of living (Atkinson, 2000; Wilhelmsson et al., 2021), thus forcing poor households to migrate or rendering them homeless.

In the energy sector, the increasing urban population and subsequent increase in the number of devices and infrastructures requiring regular power supply have prompted an increased demand for energy. This in turn has stimulated an increase in production, where it is reported that most of the energy (approximately 80 per cent) is generated from non-renewable sources (fossil fuels) (IEA, 2021), thus leading to increased emissions in cities.

The majority of the aforementioned challenges could be solved by adopting data-driven approaches. However, as noted in the previous section, this has not fully materialised as there are also challenges with data collection, storage, computation and analysis. Most importantly, challenges in the management and deployment of data have been spearheaded by third parties that are contracted to assist local governments. With a sizeable number of those corporations seeking to maximise their revenue streams, a majority of them are reported to deploy strategies that hamper open access to data, thus disadvantaging startups and third parties also involved in providing digital solutions (Engler, 2020). The lack of restricted access to open data blurs the transparency and integrity aspect of projects being undertaken in the public sector and this could lead to apprehension, especially from local residents. Additionally, when urban public data are not accessible to all participants, some of the benefits, like real-time responses to emerging issues, would be impossible, and this might have negative outcomes on innovation and unlocking the economic value of different technologies (Howells, 2018).

Cities can maximise the availability of data by local governments starting to invest in building their own capacities in terms of infrastructure (both soft and hard), and in training and/or recruiting a labour force that is well versed in data management dynamics (Kingsley et al., 2013). Thus, cities could increase their potential to derive the maximum benefits associated with the capacity to handle their own data. For instance, the issue of cost for handling data is seen to be relatively fair when this is done locally without involving third parties (OECD, 2019). This is particularly important as most cities, especially those in developing countries, have no capacity to generate sufficient revenues to support all their public investment plans, hence being able to reduce some overheads would be a positive outcome (Kingsley et al., 2013). It is reported that cities in Pakistan only manage to generate 7 per cent of their income from direct revenue streams, hence they have no capacity to finance investment projects from public budgets (Hamilton & Zhu, 2017). Building their own capacity would allow governments to take control of the quality of data to draw insights and inform decisions on diverse urban parameters. Local governments would also be able to ensure that sensitive data are handled in a way that does not compromise on the privacy and security of its residents, which is not the case when contracted private players are in control as they have profit-making agendas that prompt the monetisation of data. While it is true that it would be hard for local governments to handle all data management services by themselves, having their own capacity would allow for a better-informed public-private partnership model, where the private sector can be contracted to provide specialised services on technical aspects while local governments manage the other elements. Local governments taking the reins of data management have the potential to promote public participation, especially in avenues like civic education where citizens are actively informed of the reason for data collection for a particular project during its early stages (Bingöl, 2022; Tan & Taeihagh, 2020). This approach does not only promote project acceptability but also enhances trust which is critical in ensuring that data privacy and security concerns are alleviated. Transparency on the purpose of data – how it is being collected, stored and used – further helps in improving the quality and reliability of the data; hence, the resulting decisions and insights have far-reaching positive impacts.

WHO TAKES WHAT ROLE?

Modern cities have become attuned to the use of data gathered from vast urban components and fabrics in decision making. This was evident from mid-2000 when the uptake of the smart city planning model started to gain traction in most cities. The processes of data manipulation were spearheaded by urban managers in their quest to explore new ways of addressing endemic urban challenges. However, despite the visible positive outcomes that have been achieved in areas like alternative energy (Ferroukhi et al., 2016; Perea-Moreno et al., 2018; Phuangpornpitak & Tia, 2013), alternative mobility options (Guislain & Dasgupta, 2015; Kopestinsky, 2021; Porru et al., 2020), the adoption of sustainable practices in areas like construction and others (Deloitte, 2021; IEA & UNEP, 2018), concerns of data privacy and security that demand the attention of all stakeholders arose. In this case, national and local governments, the private sector and urban residents have direct and/or indirect roles to play to ensure that data usage in urban areas is maximised and that negative impacts are minimised. A majority of national governments have been at the forefront of formulating and instituting laws and policies that are to be followed with regard to data management within their jurisdictions. However, even before coming to the level of national governments, there are global and regional legal frameworks that have been formulated in different areas to help individual governments take control of data usage within their boundaries. For instance, in Europe, the role of data protection is guided by laws and regulations formulated at the European Union level. This began as early as 1995, when the European Union came up with the Data Protection Directive covering all the member states (Hoofnagle et al., 2019). This was later upgraded in 2018 and the directive rechristened as the General Data Protection Regulation (GDPR) (United Nations Conference on Trade and Development, 2016). In addition to this law, individual governments have the mandate to use regulations to formulate data protection laws that would subsequently guide their local governments. However, regarding smart cities, there is an array of platforms where data are transmitted to and from databases that are in most cases controlled by large corporations such as IBM and Cisco (M. Z. Allam, 2018; Allam & Newman, 2018). These cannot be localised but they too need to comply with local laws and regulations.

National governments therefore are responsible for the enactment of data laws, frameworks and guidance. As a result, the world is amass with numerous laws and guidelines. The end game for many of those laws has been to ensure that data collection, storage and use at whatever level are done in a manner that is strictly limited to the purpose for which the data were collected (United Nations Conference on Trade and Development, 2016). Governments are also key in ensuring that the processes adopted in data collection, storage and use are within the confines of the law; that they are accurate and promote fairness and transparency. They should ensure that there are enough accountable agencies or entities that guide and monitor adherence to set laws and regulations. In addition, national governments are responsible for providing and ensuring the security of physical infrastructures and installations is guaranteed.

At the local level, municipal authorities play a critical role in facilitating and enhancing data protection within their jurisdictions. The role of local governments in particular is to ensure that transparency is maintained in all matters that directly affect the public, especially with regard to smart city projects, as well as guaranteeing the privacy and security of both its citizens and that of the private sector (Macmanus et al., 2013). These two roles are sometimes argued to be conflicting as they demand that local governments maintain an intricate balance between the two. Without that balance there may arise concerns of data access control, which can result in restrictions on access to public assets. On the other hand, when open access is explicitly provided for, the probability of a lack of privacy concerns becomes very high. This then behoves local governments to also consult regulations and guidelines, provided in the GDPR, as well as other localised laws (United Nations Conference on Trade and Development, 2016). These regulations are critical at local governance levels as they help foster a wide range of digital innovations and solutions to different urban challenges. In order to effectively comply with established regulations, local governments have the responsibility of ensuring that sufficient budgetary allocations are made on data protection. Local governments, especially in developing and less developed economies, might not have the prerequisite financial capacities to effectively guarantee sufficient financial resources (Pisa et al., 2021). They then need to liaise with national governments, the private sector or other financial partners to ensure that sufficient funds are available. In addition, as noted in the previous section, local governments need to build their own capacity to undertake some, if not all, the data management responsibility. This would entail equipping part of its labour force with the technical skills and expertise that would allow them to work within data management systems, including in legal departments, to be able to interpret and enforce global, regional and national data protection laws governing collection, storage and use (Pisa et al., 2021). This is important as most of the private firms and companies interested in data collection, storage and use might have extensive technical expertise, hence local governments need to be equally well equipped so that they are able to collaborate cohesively.

The private sector, especially those directly involved at various levels of data management, also has a significant role to play to ensure that they do not compromise or water down established privacy and security protection measures. In most cases, especially in the situation of smart cities, the private sector is considered a major partner in the implementation of projects, providing the technologies, digital products and expertise that most local governments lack (Kingsley et al., 2013). However, they are also motivated in their work by the prospect of making substantial profit and this might impact their attitude to data handling and management. Therefore, to ensure coherence with their work and the data protection agenda set out in local, national and regional guidelines, the private sector needs to be actively involved in co-creating guidelines and codes of practice (United Nations Conference on Trade and Development, 2016). These would not only help them remain within established rules, but would also allow regulators (being local governments) to introduce customised regulations that match local conditions and demands.

Residents, too, are an important constituent in data management agendas. In fact, they are the most vulnerable when there is a breach or security threat on available data. While urban data are gathered from different urban elements, a sizeable amount is directly collected from residents, thus it often contains sensitive personal identifying information that needs to be closely guarded. In some instances, some of the personal data are collected without the knowledge or consent of the residents, infringing on their privacy and rights. Such issues, as prescribed in the GDPR, need to be ironed out by actively engaging the public on ongoing or intended objectives of different urban projects, especially the need for data. This ensures that residents are aware that their data are being collected and are also informed of the intended use of the same data. When such participatory platforms are available, residents can be more involved and suggest ways their data could be handled. Similarly, they are able to get clarification on how their data are being protected. With this information, they are further able to pursue the right channels if they feel the privacy and security of their personal information is being threatened by different activities taking place in cities.

REFERENCES

Afrin, T., & Yodo, N. (2020). A survey of road traffic congestion measures towards a sustainable and resilient transportation system. *Sustainability*, *12*(11), 4660.

Alenezi, M. (2021). Safeguarding cloud computing infrastructure: A security analysis. *Computer Systems Science and Engineering*, *37*, 159–167.

Allam, M. Z. (2018). Redefining the smart city: Culture, metabolism and governance. Case study of Port Louis, Mauritius. Curtin University. Retrieved 1 February 2022 from https://espace.curtin.edu.au/handle/20.500.11937/70707

Allam, Z. (2018). On smart contracts and organisational performance: A review of smart contracts through the blockchain technology. *Review of Economic Business Studies*, *11*(2), 137–156.

Allam, Z. (2019a). *Cities and the Digital Revolution: Aligning Technology and Humanity*. Springer Nature.

Allam, Z. (2019b). The emergence of anti-privacy and control at the nexus between the concepts of safe city and smart city. *Smart Cities*, *2*(1), 96–105.

Allam, Z. (2020a). Biometrics, privacy, safety, and resilience in future cities. In Z. Allam (Ed.), *Biotechnology and Future Cities: Towards Sustainability, Resilience and Living Urban Organisms* (pp. 69–87). Springer International Publishing.

Allam, Z. (2020b). Data as the new driving gears of urbanization. In *Cities and the Digital Revolution* (pp. 1–29). Palgrave Pivot.

Allam, Z. (2020c). The first 50 days of COVID-19: A detailed chronological timeline and extensive review of literature documenting the pandemic. *Surveying the COVID-19 Pandemic and Its Implications*, *1*, 1–7.

Allam, Z. (2020d). The rise of machine intelligence in the COVID-19 pandemic and its impact on health policy. *Surveying the COVID-19 Pandemic and Its Implications*, *89*.

Allam, Z. (2020e). The second 50 days: A detailed chronological timeline and extensive review of literature documenting the COVID-19 pandemic from day 50 to day 100. *Surveying the COVID-19 Pandemic and Its Implications*, *9*.

Allam, Z. (2021). Big data, artificial intelligence and the rise of autonomous smart cities. In *The Rise of Autonomous Smart Cities* (pp. 7–30). Palgrave Macmillan.

Allam, Z., & Dhunny, Z. A. (2019). On big data, artificial intelligence and smart cities. *Cities*, *89*, 80–91.

Allam, Z., & Jones, D. S. (2020). On the Coronavirus (COVID-19) outbreak and the smart city network: Universal data sharing standards coupled with artificial intelligence (AI) to benefit urban health monitoring and management. *Healthcare*, *8*(1), 46.

Allam, Z., & Newman, P. (2018). Redefining the smart city: Culture, metabolism and governance. *Smart Cities*, *1*(1). https://doi.org/10.3390/smartcities1010002

Allam, Z., Tegally, H., & Thondoo, M. J. S. C. (2019). Redefining the use of big data in urban health for increased liveability in smart cities. *Smart Cities*, *2*(2), 259–268.

Armstrong, M. (2021, 24 August). The world's safest cities. Statista. Retrieved 6 March 2022 from www.statista.com/chart/3178/the-worlds-safest-cities/

Atkinson, R. (2000). The hidden costs of gentrification: Displacement in central London. *Journal of Housing and the Built Environment*, *15*(4), 307–326.

Azodo, I., Williams, R., Sheikh, A., & Cresswell, K. (2020). Opportunities and challenges surrounding the use of data from wearable sensor devices in health care: Qualitative interview study. *Journal of Medical Internet Research*, *22*(10), e19542–e19542.

Barrios, S., Bertinelli, L., & Strobl, E. (2006). Climate change and rural–urban migration: The case of sub-Saharan Africa. *Journal of Urban Economics*, *60*(3), 357–371.

Bezgrebelna, M., McKenzie, K., Wells, S., Ravindran, A., Kral, M., Christensen, J., Stergiopoulos, V., Gaetz, S., & Kidd, S. A. (2021). Climate change, weather, housing precarity, and homelessness: A systematic review of reviews. *International Journal of Environmental Research and Public Health*, *18*(11). https://doi.org/10.3390/ijerph18115812

Bibri, S. E. (2020). Data-driven environmental solutions for smart sustainable cities: Strategies and pathways for energy efficiency and pollution reduction. *Euro-Mediterranean Journal for Environmental Integration*, *5*(3), 66.

Bibri, S. E. (2021a). Data-driven smart eco-cities and sustainable integrated districts: A best-evidence synthesis approach to an extensive literature review. *European Journal of Futures Research*, *9*(1), 16.

Bibri, S. E. (2021b). Data-driven smart sustainable cities of the future: An evidence synthesis approach to a comprehensive state-of-the-art literature review. *Sustainable Futures*, *3*, 100047.

Bibri, S. E. (2021c). Data-driven smart sustainable cities of the future: Urban computing and intelligence for strategic, short-term, and joined-up planning. *Computational Urban Science*, *1*(1), 8.

Bingöl, E. S. (2022). Citizen participation in smart sustainable cities. In E. A. Babaoğlu & O. Kulaç (Eds), *Research Anthology on Citizen Engagement and Activism for Social Change* (pp. 967–987). IGI Global.

Bond, D., & Kynge, J. (2018, 3 December). Huawei under fire as politician fret over 5G security. *Financial Times*. Retrieved 13 October 2019 from www.ft.com/content/8bb75604-f4a6-11e8-ae55-df4bf40f9d0d

Bouton, S., Hannon, E., Knupfer, S., & Ramkumar, S. (2017, June). The future (s) of mobility: How cities can benefit. McKinsey. Retrieved 15 January 2020 from www.mckinsey.com/business-functions/sustainability/our-insights/the-futures-of-mobility-how-cities-can-benefit

Brasuell, J. (2015, 22 June 2015). The early history of the 'smart cities' movement – in 1974 Los Angeles. Planetizen. Retrieved 20 January 2021 from www.planetizen.com/node/78847

Bulao, J. (2022, 6 February 2022). How much data is created every day in 2022? Tech Jury. Retrieved 7 March 2022 from https://techjury.net/blog/how-much-data-is-created-every-day/#gref

Chamie, J. (2017, 13 July). As cities grow, so do the numbers of homeless. Yale Global. Retrieved 3 February 2022 from https://archive-yaleglobal.yale.edu/content/cities-grow-so-do-numbers-homeless

Cipriani, J. (2021, 9 September). T-Mobile data breach 2021: Here's what it means for securing your data. Cnet. Retrieved 6 March 2022 from www.cnet.com/how-to/t-mobile-data-breach-2021-heres-what-it-means-for-securing-your-data/

Cisco. (2015, 28 October). Cisco global cloud index projects cloud traffic to quadruple by 2019. Cisco. Retrieved 8 March 2022 from https://newsroom.cisco.com/press-release-content?articleId=1724918

Cisco. (2016). Digital transformation for public safety: Enables dynamic operational environments. Cisco. Retrieved 7 March 2022 from www.cisco.com/c/dam/en_us/solutions/industries/docs/gov/digital-transformation-public-safety.pdf

Davis, J. (2020, 6 April). FBI: COVID-19 spurs increase in Zoom, video-conferencing hijacking. Health IT Security. Retrieved 8 March 2022 from https://healthitsecurity.com/news/fbi-covid-19-spurs-increase-in-zoom-video-conferencing-hijacking

Dawar, N., & Frost, T. (1999, April). Competing with giants: Survival strategies for local companies in emerging markets. *Harvard Business Review*. Retrieved 8 March 2022 from https://hbr.org/1999/03/competing-with-giants-survival-strategies-for-local-companies-in-emerging-markets

Deloitte. (2021). 2021 engineering and construction industry outlook. Deloitte. Retrieved 25 August 2021 from www2.deloitte.com/us/en/pages/energy-and-resources/articles/engineering-and-construction-industry-trends.html

Depoy, J., Phelan, J., Sholander, P., Smith, B., Varnado, G. B., & Wyss, G. (2005, 17–20 October). Risk assessment for physical and cyber attacks on critical infrastructures. MILCOM 2005 IEEE Military Communications Conference.

Desjardins, J. (2019, 17 April). How much data is generated each day? World Economic Forum. Retrieved 8 March 2022 from www.weforum.org/agenda/2019/04/how-much-data-is-generated-each-day-cf4bddf29f/

Economist Intelligence Unit. (2021). Safe cities index 2021. Economist Intelligence Unit. Retrieved 6 March 2022 from https://safecities.economist.com/

Engler, A. (2020, 10 September). Tech cannot be governed without access to its data. Brooking. Retrieved 8 March 2022 from www.brookings.edu/blog/techtank/2020/09/10/tech-cannot-be-governed-without-access-to-its-data/

European Automobile Manufacturers Association. (2021). Vehicle in use Europe. Report, ACEA. Retrieved 2 February 2022 from www.acea.auto/files/report-vehicles-in-use-europe-january-2021-1.pdf

European Commission, Eurostat, & DG. (2020). A recommendation on the method to delineate cities, urban and rural areas for international statistical comparisons. Retrieved from https://unstats.un.org/unsd/statcom/51st-session/documents/BG-Item3j-Recommendation-E.pdf

Ferroukhi, R., Lopez-Peña, A., Kieffer, G., Nagpal, D., Hawila, D., Arslan Khalid, El-Katiri, L., Vinci, S., & Fernandez, A. (2016). Renewable energy benefits: Measuring the economics. IRENA. Retrieved from www.irena.org/-/media/Files/IRENA/Agency/Publication/2016/IRENA_Measuring-the-Economics_2016.pdf

Ghafir, I., Saleem, J., Hammoudeh, M., Faour, H., Prenosil, V., Jaf, S., Jabbar, S., & Baker, T. (2018). Security threats to critical infrastructure: The human factor. *Journal of Supercomputing*, *74*(10), 4986–5002.

Giyasov, B., & Giyasova, I. (2018). High-rise construction 2017 (HRC 2017). *Architecture an Urban Planning*, *33*, 7. https://doi.org/10.1051/e3sconf/20183301050

Glaeser, E. L., & Shapiro, J. M. (2001). *Cities and Warfare: The Impact of Terrorism on Urban Form*. Harvard University. Retrieved 2 February 2022 from www.brown.edu/Research/Shapiro/pdfs/HIER1942.pdf

Granville, K. (2018, 19 March). Facebook and Cambridge Analytica: What you need to know as fallout widens. *New York Times*. Retrieved 6 March 2022 from www.nytimes.com/2018/03/19/technology/facebook-cambridge-analytica-explained.html

Guislain, P., & Dasgupta, A. (2015, 15 January). Who needs cars? Smart mobility can make cities sustainable. World Bank. Retrieved 10 December 2021 from https://blogs.worldbank.org/transport/who-needs-cars-smart-mobility-can-make-cities-sustainable

Hamilton, S., & Zhu, X. (2017). Funding and financing smart cities. Retrieved 2 February 2022 from www2.deloitte.com/content/dam/Deloitte/us/Documents/public-sector/us-ps-funding-and-financing-smart-cities.pdf

Help Net Security. (2020, 3 April). While nearly 90% of companies are backing up data, only 41% do it daily. Help Net Security. Retrieved 7 March 2022 from www.helpnetsecurity.com/2020/04/03/back-up-data/

Holmes, A. (2021, 3 April). 533 million Facebook users' phone numbers and personal data have been leaked online. *Business Insider Africa*. Retrieved 6 Apr 2022 from https://africa.businessinsider.com/tech/533-million-facebook-users-phone-numbers-and-personal-data-have-been-leaked-online/65sy87q

Hoofnagle, C. J., van der Sloot, B., & Borgesius, F. Z. (2019). The European Union general data protection regulation: What it is and what it means. *Information and Communications Technology Law*, *28*(1), 65–98.

Howells, J. (2018, 12 March). Smart cities need open data. Orange Business. Retrieved 8 March 2022 from www.orange-business.com/en/blogs/smart-cities-need-open-data

IEA. (2021). Electricity explained: Electricity in the United States. EIA. Retrieved 26 September 2021 from www.eia.gov/energyexplained/electricity/electricity-in-the-us.php

IEA, & UNEP. (2018). 2018 Global status report: Towards a zero-emission, efficient and resilient buildings and construction sector.

International Organization for Migration. (2019). *Climate Change and Migration in Vulnerable Countries: A Snapshot of Least Developed Countries, Landlocked Developing Countries and Small Island Developing States*. International Organization for Migration.

Ismagilova, E., Hughes, L., Rana, N. P., & Dwivedi, Y. K. (2020). Security, privacy and risks within smart cities: Literature review and development of a smart city interaction framework. *Information Systems Frontiers*. https://doi.org/10.1007/s10796-020-10044-1

Jibilian, I., & Canales, K. (2021, 15 April). The US is readying sanctions against Russia over the SolarWinds cyber attack. Here's a simple explanation of how the massive hack happened and why it's such a big deal. *Business Insider*. Retrieved 6 March 2022 from www.businessinsider.com/solarwinds-hack-explained-government-agencies-cyber-security-2020-12?r=US&IR=T

Kelman, I. (2015). Difficult decisions: Migration from small island developing states under climate change. *Earth's Future*, *3*(4), 133–142.

Kingsley, G. T., Pettit, K. L. S., & Hendey, L. (2013). Strengthening local capacity for data-driven decision making. Retrieved 2 February 2022 from www.urban.org/sites/default/files/publication/23901/412883-Strengthening-Local-Capacity-For-Data-Driven-Decisionmaking.PDF

Kopestinsky, A. (2021, 12 August). Electric car statistics in the US and abroad. Policy Advice. Retrieved 10 December 2021 from https://policyadvice.net/insurance/insights/electric-car-statistics/#:~:text=Well%2C%20the%20latest%20figures%20show,3.4%20million%20to%205.6%20million

Koretsky, Z., & van Lente, H. (2020). Technology phase-out as unravelling of socio-technical configurations: Cloud seeding case. *Environmental Innovation and Societal Transitions*, *37*, 302–317.

Kroposki, B., Johnson, B., Zhang, Y., Gevorgian, V., Denholm, P., Hodge, B., & Hannegan, B. (2018). Achieving a 100% renewable grid: Operating electric power systems with extremely high levels of variable renewable energy. *IEEE Power and Energy Magazine*, *15*(2), 61–73.

Macmanus, S. A., Caruson, K., & McPhee, B. D. (2013). Cybersecurity at the local government level: Balancing demands for transparency and privacy rights. *Journal of Urban Affairs, 35*(4), 451–470.

Maitre, N., Montt, G., Saget, C., Ernst, C., Gutiérrez, M. T., Kizu, T., Karimova, T., Lieuw-Kie-Song, M., & Tsukamoto, M. (2018). The employment impact of climate change adaptation: Input Document for the G20 Climate Sustainability Working Group. Retrieved 1 February 2022 from www.ilo.org/wcmsp5/ groups/public/---ed_emp/documents/publication/wcms_645572.pdf

Mascitelli, B., & Chung, M. (2019). Hue and cry over Huawei: Cold war tensions, security threats or anti-competitive behaviour? *Research in Globalization, 1*, 100002.

Mitchell, K. C. (2021, 15 November). Training a resilient workforce to secure our critical infrastructure. US Department of Homeland Security, Science and Technology Directorate. Retrieved 7 March 2022 from www.dhs.gov/ science-and-technology/news/2021/11/15/training-resilient-workforce-secure -our-critical-infrastructure

Mordor Intelligence. (2021). Smart cities market: Growth, trends, COVID-19 impact, and forecasts (2022–2027). Mordor Intelligence. Retrieved 8 March 2022 from www.mordorintelligence.com/industry-reports/smart-cities-market #:~:text=The%20smart%20cities%20market%20was,market%20in%20the %20forecast%20period

Newman, L. H. (2018, 23 April). Atlanta spent $2.6m to recover from a $52,000 Ransomware scare. Wired. Retrieved 8 March 2022 from www.wired.com/ story/atlanta-spent-26m-recover-from-ransomware-scare/

OECD. (2019). Enhancing the contribution of digitalisation to the smart cities of the future. OECD. Retrieved 2 February 2022 from www.oecd.org/cfe/ regionaldevelopment/Smart-Cities-FINAL.pdf

OECD, European Commission, FAO, UN-Habitat, World Bank, & Eurostat. (2021). Applying the degree of urbanisation. https://doi.org/doi:https://doi .org/10.1787/4bc1c502-en

Partida, D. (2021, 26 September). The physical security challenges of smart cities. Planetizen. Retrieved 7 March 2022 from www.planetizen.com/blogs/ 114762-physical-security-challenges-smart-cities

Perea-Moreno, M., Hernandez-Escobedo, Q., & Perea-Moreno, A. (2018). Renewable energy in urban areas: Worldwide research trends. *Energies, 11*, 577.

Phuangpornpitak, N., & Tia, S. (2013). Opportunities and challenges of integrating renewable energy in smart grid system. *Energy Procedia, 34*, 282–290.

Pisa, M., Dixon, P., & Nwankwo, U. (2021, 6 December). Why data protection matters for development: The case for strengthening inclusion and regulatory capacity. Center for Global Development. Retrieved 9 March 2022 from www.cgdev.org/publication/why-data-protection-matters-development-case -strengthening-inclusion-and

Platt, F. (2012). Physical threats to the information infrastructure. In S. Bosworth, M. E. Kabay, & E. Whyne (Eds), *Computer Security Handbook* (pp. 22.21–22.28). Wiley.

Porru, S., Misso, F. E., Pani, F. E., & Repetto, C. (2020). Smart mobility and public transport: Opportunities and challenges in rural and urban areas. *Journal of Traffic and Transportation Engineering, 7*(1), 88–97.

Rodríguez-Pose, A., & von Berlepsch, V. (2019). Does population diversity matter for economic development in the very long term? Historic migration, diversity and county wealth in the US. *European Journal of Population, 35*(5), 873–911.

See, A. v. (2021, 7 June). Amount of data created, consumed, and stored 2010–2025. Statista. Retrieved 26 January 2022 from www.statista.com/statistics/871513/worldwide-data-created/#:~:text=Over%20the%20next%20five%20years,to%20more%20than%20180%20zettabytes

Statista Research Department. (2016, 27 November). Internet of Things – number of connected devices worldwide 2015–2025. Statista. Retrieved 13 December 2019 from www.statista.com/statistics/471264/iot-number-of-connected-devices-worldwide/

Strengholt, P. (2020). *Data Management at Scale*. O'Reilly Media.

Strinsjö, J. (2020, 24 November). The smart city: Why the silo mindset will limit its full potential. PBC Today. Retrieved 8 March 2022 from www.pbctoday.co.uk/news/bim-news/smart-city-potential/85685/

Tan, S. Y., & Taeihagh, A. (2020). Smart city governance in developing countries: A systematic literature review. *Sustainability, 12*(3). https://doi.org/10.3390/su12030899

Thenuan, P., & Raina, A. (2016). Data privacy concerns for big data monetisation in mobile application. *Prayukti-Student Research Journal*. Retrieved from www.researchgate.net/publication/309012928_DATA_PRIVACY_CONCERNS_FOR_BIG_DATA_MONETIZATION_IN_MOBILE_APPLICATION

UN. (2019). World population prospects 2019. Retrieved 2 February 2022 from https://population.un.org/wpp/Publications/Files/WPP2019_DataBooklet.pdf

UNDESA. (2018, 16 May). 68% of the world population projected to live in urban areas by 2050, says UN. United Nations. Retrieved 5 January 2021 from www.un.org/development/desa/en/news/population/2018-revision-of-world-urbanization-prospects.html

UNDESA. (2021). Goal 11: Make cities and human settlements inclusive, safe, resilient and sustainable. UNDESA. Retrieved 4 February 2022 from https://sdgs.un.org/goals/goal11

United Nations. (2011). *Transforming our World: The 2030 Agenda for Sustainable Development* (A/RES/70/1). Retrieved 2 February 2022 from https://sustainabledevelopment.un.org/content/documents/21252030%20Agenda%20for%20Sustainable%20Development%20web.pdf

United Nations. (2017). *New Urban Agenda*. H. I. Secretariat. Retrieved from https://uploads.habitat3.org/hb3/NUA-English.pdf

United Nations Conference on Trade and Development. (2016). *Data Protection Regulations and International Data Flows: Implications for Trade and Development*. Retrieved from https://unctad.org/system/files/official-document/dtlstict2016d1_en.pdf

Wilhelmsson, M., Ismail, M., & Warsame, A. (2021). Gentrification effects on housing prices in neighbouring areas. *International Journal of Housing Markets and Analysis*. https://doi.org/10.1108/IJHMA-04-2021-0049

World Bank. (2020, 20 April). *Urban Development: Overview*. World Bank. Retrieved 27 September 2021 from www.worldbank.org/en/topic/urbandevelopment/overview#:~:text=With%20more%20than%2080%25%20of,and%20new%20ideas%20to%20emerge

Yigitcanlar, T., Kamruzzaman, M., Foth, M., Sabatini-Marques, J., da Costa, E., & Ioppolo, G. (2019). Can cities become smart without being sustainable? A systematic review of the literature. *Sustainable Cities and Society*, *45*, 348–365.

Yu, Z., Zhang, H., Tao, Z., & Liang, J. (2019). Amenities, economic opportunities and patterns of migration at the city level in China. *Asian and Pacific Migration Journal*, *28*(1), 3–27.

Zarifis, A., Kawalek, P., & Azadegan, A. (2021). Evaluating if trust and personal information privacy concerns are barriers to using health insurance that explicitly utilizes AI. *Journal of Internet Commerce*, *20*(1), 66–83.

Zetter, K. (2016, 3 March). Inside the cunning, unprecedented hack of Ukraine's power grid. Wired. Retrieved 8 March 2022 from www.wired.com/2016/03/inside-cunning-unprecedented-hack-ukraines-power-grid/

Zoonen, L. v. (2016). Privacy concerns in smart cities. *Government Information Quarterly*, *33*(3), 472–480.

6. Enter 6G and the augmented smart city

INTRODUCTION

Cities have grown and benefitted immensely since the advent and subsequent widespread use of technologies – even before the emergence of the Industrial Revolution. During the Mesopotamian era, for instance, there were technologies in areas like glassmaking, metalworking, irrigation and textile weaving. This prompted the emergence and exponential growth of many cities in ancient civilisations (Berger et al., 2016). The advent of the Industrial Revolution raised the bar for cities even higher and from a historical point of view cities have managed to remain abreast with varying periods of industrial revolutions. It is reported that the population of England and Wales, which were among the first beneficiaries of the Industrial Revolution, grew dramatically in 1879, prompted by a perceived increase in quality of life. However, during this period, only 10 per cent of the population globally lived in urban areas (Britannica, 2020). During the third industrial revolution, which began in the late 1900s characterised by extensive digitisation and automation, cities experienced massive transformation, including an increased use of automobiles, electronic devices, computers and the internet (Ward, 2019). Such growth in urban areas was stimulated by the rapid urban population growth. In the 1960s, the ratio of urban–rural population was estimated to be approximate 1:3. However, by 2007, the ratio was at equilibrium (3.33 billion people in each setting) (Ritchie & Roser, 2019). By 2015, more than 54 per cent of the urban population were living in urban areas (Kaneda et al., 2020).

The rapid transformation in urban population and the subsequent urbanisation are credited to the emergence of the fourth industrial revolution in the mid-2000s where the use of information and communication technology (ICT) became the norm in most cities. The use of the internet become robust, supporting communication within different platforms and allowing for globalisation that stimulated further growth for cities. As a result, massive attention and effort has been directed into ICT usage as is evidenced by the transformation in broadband connectivity speed. This began in the form of the General Packet Radio Service, which then transformed to 2G, 3G and then 4G, each emerging at an interval of ten years from the 1970s. The latest is 5G (launched in 2019), which has gained massive traction in most major cities across the globe. 5G technology is now favoured in most cities as it allows for massive transfer of data in real time and supports flawless communication between different smart devices installed in cities due to its relatively low latency compared to its predecessors (Oughton et al., 2019). However, even before different transformations in mobile communication networks came about, the mere presence of communication capabilities via this network benefitted cities in diverse ways. These include allowing for digitalisation and creation opportunities for the emergence of new job opportunities through increased economic scope. They also stimulated innovations in sectors like culture and art where it became possible for cities to start preserving and conserving some of their priceless heritage in the virtual environment (Di Giulio et al., 2019). Currently, with 4G and 5G, the capabilities in cities have become widespread and robust in different sectors such as transport, health, finance, housing, energy, environment, education and entertainment.

In particular, 5G technology has been hailed for the potential it brings to cities, especially in allowing for massive connection of various devices and the subsequent sharing of data. This is critical for smart cities as the number of Internet of Things (IoT) smart devices are currently on the rise, and with the aftermath of COVID-19 that prompted an increased attention on digital solutions, it is expected that connected devices will continue to grow, projected to reach 75 billion connected devices by 2025 (Statista Research Department, 2016). These devices will need to share data in real time and in a complex network which is not sufficiently supported by 4G technology, hence the rapid uptake of 5G. For instance, post-pandemic, there was anticipation of a *metaverse* ecosystem – a virtual reality environment that mirrors the physical world allowing people to interact, socialise, work and do almost everything in the virtual realm. This will entail close collaboration and interaction between different existing and emerging technologies such as digital twins, artificial intelligence (AI), cloud computing and augmented reality (AR) etc. Further, even before the world gets to experience this immersive experience, there are some realities like working from home, virtual communications (e.g. teleconferencing), cashless ecosystems and cryptocurrency technologies that require high-speed communication and warrant the existence of 5G technology (Yang et al., 2022).

However, the prospects and strides already made possible by the existence of 5G technology are still not sufficient to power all the digital demands. For instance, some stakeholders in the technology world still require a mobile communication network that will guarantee lower latency than currently provided, higher communication speed and higher frequencies (Nawaz et al., 2020). These might be addressed in the antic-ipated 6G cellular technology that some ICT giants such as Samsung are said to be exploring (Samsung, 2020). The objective of Samsung and other ICT firms active in this quest is to ensure that clients experience hyperconnectivity capabilities, allowing humans to have total control of the world around them as well as control their virtual worlds in a seam-less manner.

For smart cities, the existence of 5G technology seems to be enough to pursue objectives such as sustainability agendas and socioeconomic development. For instance, with this technology, there are prospects of improving sectors like cultural heritage and art, smart tourism projects and improving capabilities in the health, sport and education sectors (Rong et al., 2020). 6G is anticipated to bring a new experience altogether in these sectors, including allowing the emergence of new technologies such as comprehensive AI, split computing and duplex technologies, as will be discussed below (Allam & Jones, 2021).

When coupled with the already existing and advancing technology of AR, the 6G technology is expected to transform urban areas, including existing smart cities, to allow them to achieve true 'smartness'. In light of the above background, this chapter explores how 6G technology might operate in smart cities, especially its impact on existing technologies that support and allow for data generation, storage, computation and use.

FROM 5G TO 6G, AND THE IMPACT OF THE INTERNET OF THINGS

Although 5G and 6G will take some years to further penetrate the market, it is expected that they will prompt an emergence of new tech-nologies that will make the smart city concept even more achievable. They are expected to allow for an unlimited number of IoT devices to be plugged into the physical urban infrastructure without the fear of data management challenges that exists now. The first generation (1G) of mobile wireless connectivity emerged in the 1970s and lasted until 1994, when 2G emerged (International Telecommunication, 2011). 1G boasted a frequency of only 30 Khz and a bandwidth of only 2 kilobytes per second. Ten years later, even before 1G was phased out, 2G, with an increased frequency and bandwidths of 1.8 Ghz and 14.4–64 kbps, emerged. This was succeeded 10 years later (2001) by 3G, which had an improved frequency of approximately 2 Ghz and a bandwidth of approximately 2 Mbps. It was powered by the Packet Network as the core network – which is a departure from its predecessors that rode on the Public Switched Telephone Network (Galazzo, 2020). In the wake of the twenty-first century (in 2010), 4G was introduced boasting an increased bandwidth of approximately 200 Mbps–1 Gbps and a frequency of approximately 2 Mbps and running on the internet as its core network (Akhtar, 2005; Khan et al., 2009). Its emergence brought about a paradigm shift in the wireless connectivity field, as the frequency and impressive bandwidth allowed internet users to improve their capacity, especially in data collection and sharing. It allowed and prompted the emergence of numerous ground-breaking digital innovations such as ride sharing, e-commerce models, remote working and other frontiers that changed social, economic and sustainability dimensions (Moore, 2020). However, 5G, introduced in 2016 (and launched in 2019), changed the mobile connectivity terrain in a tremendous way. This allowed for an increased frequency rate of between 3 Ghz and 30 Ghz and also increased the connectivity speed by more than 1 Gbps, reaching a high of between 15 to 20 Gbps in some areas (Guilford, 2019).

In fact, the potential of 5G attracted some criticism, especially due to its prospective connectivity speed. In particular, some governments feared it would escalate insecurity issues in cities (Ahmad et al., 2017). A classical case is its incorporation in smartphone devices by Huawei – a move that was interpreted as a strategy by the Chinese government to enable espionage (Mascitelli & Chung, 2019). However, beyond geopolitical concerns, the potential of 5G, especially in unlocking the full potential of smart cities, is unmatched for now. For instance, regarding data transfer, experts believe that 5G is unlocking the true real-time transfer of data between billions of installed IoT devices and central networks (Rong et al., 2020). This will help solve some critical issues such as connectivity speed and data storage problems which are viewed as the main hurdles in the smart city concept implementation. This happens as 5G allows for the quick transfer of data into cloud data management platforms. In addition to the quick transfer of data, 5G is seen to have bolstered the smart city market exponentially. For instance, since the emergence of 5G, the market valuation for smart cities is expected to reach a high of approximately $2.5 trillion by 2026 (Globenewswire, 2022). On its part, the 5G technology was valued at $69.93 billion in 2020 and reached a high of $83.24 billion in 2021, at a compounded annual growth rate of 20 per cent (Reach and Markets, 2021). Such statistics ring true despite the fact that 5G has not been deployed extensively in most cities, especially in the global south, due to the limitation of supportive infrastructures (Rahman et al., 2021). However, efforts to make it the mainstream mobile connectivity network are evident, spearheaded in particular by mobile phone manufacturers and mobile service providers.

Beyond 5G, there is anticipation that the 6G cellular technology might be in the offing in the near future, with an earliest expected launch date of 2030 as per a white paper authored by Samsung (2020). However, its success will be tied to the success of 5G, especially in relation to other technologies being deployed in smart cities. 6G will be in a league of its own when compared to its predecessors in terms of speed, latency and frequency. For instance, it is projected that the technology might achieve a frequency of over 95 Ghz to 3 Thz and a mobile bandwidth of approximately 1 Terabyte per second (>1 Tbps) (Chowdhury et al., 2019). This will be approximately 1,000 times faster than 5G, which is currently the most advanced mobile wireless connectivity technology.

When the 5G and later the 6G technologies are coupled with other smart urban technologies such as IoT, AI, digital twins and others, the anticipation is that cities will continue to experience an unprecedented increase in innovation and production. Advanced tools such as AR that will help significantly in addressing environmental sustainability concerns will henceforth become mainstream. 6G will play a significant role in natural resource conservation, especially by promoting innovation on the creation of portable geographic information systems and global positioning systems. These will help in promoting innovations in the transport sector, especially with regard to the increase of autonomous cars in cities. In other sectors such as health, proponents of this technology argue that it will enhance performance in areas like telesurgery, the emergence and use of advanced wearables and 3D printing, among other things (Nayak & Patgiri, 2020). With prospects such as the metaverse being championed as the next urban ecosystem frontier in cities, especially after the realities of COVID-19 (Theo, 2021), the capabilities of 6G will enhance the applicability of diverse technologies such as cryptocurrency and digital twins (Hall & Li, 2021). It will further make the virtual realm become completely immersive by allowing the seamless sharing of data, especially of digital replicas between the two worlds (physical and virtual).

THE RISE AND GROWTH OF AUGMENTED SMART CITIES

In the pursuit of smart cities, the main focus is often on influencing major urban dimensions serving as key pillars to different urban fabrics. These include mobility, economy, governance, environment, energy, people and living (Kaji et al., 2018). The objectives and anticipated outcomes for each of the dimensions influence the technologies to be adopted – though it has been widely argued that most technologies cut across different urban sectors, which is more of an advantage than when technologies are applied in silos (Allam, 2019a). In line with this, the rise and growth of AR is particularly seen to benefit a number of dimensions (living, mobility, economy and environment) in an extensive way (Allam, 2020c, 2021; Sharifi & Allam, 2021). In the mobility dimension, AR technologies have allowed the creation of mobile applications (APPs) that make it possible for those devices to be used for navigation purposes. Some APPs are even integrated in such a way that they allow cameras in different devices to be used as an aid in navigation (Facebook, 2021).

In the social dimension, which encompasses different aspects such as socialisation, recreation and entertainment, the rise of AR is observed to create the capacity for new experiences not possible in existing devices. For instance, people are now able to enjoy and interact with virtual objects in a form that seems more realistic. However, the anticipated metaverse dimension is expected to prompt advanced experiences where people will be able to interact and socialise in the virtual realm just like they do in the physical world (Samsung, 2020). AR technology has thus been championed for its ability to help overcome some physical limitations such as geographical distance and time that make it difficult for people to access and enjoy diverse things like recreation facilities. When coupled with advanced wireless mobile communication technologies (5G and later 6G) and digital twins – which is also gaining traction in smart cities – people are expected to have the opportunity to extend their experience even in areas like tourism, where replicas of different attraction sites and components will be readily available (Chapman, 2021). The advanced cellular connectivity is expected to allow for media streams and displays of very high quality (4 K and 8 K), thus reducing the urge to have a physical presence in areas like tourist attractions and entertainment hubs (Samsung, 2020). This will be critical for smart cities, especially in their quest to reduce the number of automobiles.

During the first quarter and part of the second quarter of 2020 (during the height of the COVID-19 pandemic), the global social fabric was greatly disrupted. For instance, over half of the global population was confined in lockdowns thus disrupting the social order (Allam, 2020a, 2020b, 2020d). It became untenable to continue with activities in sports, learning institutions, entertainment hubs and recreation centres. The challenges escalated to the manufacturing, transport and many other sectors that generally attract social and economic activities each day. However, despite the immeasurable challenges that people were undergoing, a 'silver lining' on the continuity of social interactions was made real by the existence of a wide range of social media platforms. These made it possible for people to maintain some kind of social interaction despite the physical social distances. They also made it possible for people to break the monotony and loneliness of having to remain in their homes. For people who had access to some form of AR, coupled with a high internet connectivity speed and other smart technologies, it has been found that they were better off than those without access to these (Pallavicini et al., 2021). Also, such technologies allowed people, especially with health conditions, to continue interacting with their medical caretakers or seek further assistance elsewhere, albeit online, thus preventing them from falling behind in their medication and therapy routines (Siani & Marley, 2021). In addition, the availability of these technologies is also argued to have helped a great deal in preventing, assessing and treating medical conditions, especially those associated with mental health that were on the rise during the lockdown periods, exacerbated by issues like job losses, uncertainty about the future, anxiety and forced social isolation (Sampaio et al., 2021).

Following the unprecedented urban population growth, urban managers across the globe are keen to explore new economic frontiers that could, among other things, guarantee new job opportunities and increase revenue streams. One such frontier that has not been fully exploited is the culture and creative arts industry. In a report by UNESCO (2021), it was noted that this industry contributes significant annual revenues (approximately $2,250 billion per annum) on a global scale. Exports of different products from this sector generate approximately $250 billion, and in the new future, this industry will account for approximately 10 per cent of the total global economy (UNESCO, 2017). Among innovations that will prompt the projected growth is the continued evolution and subsequent use of AR. While it is not a new thing for AR to be deployed in the cultural and creative industries, it is notable that the existence of supportive technologies such as 5G, AI, digital twins and others are expected to make this even more economically viable (Yang et al., 2022). The applicability of AR to the cultural industry is particularly expected to influence the revitalisation of traditional heritage, artefacts and other cultural products (Charr, 2020), making them accessible to a wider audience. This will not only help in their conservation, but will also make them a source of revenue for a substantial global population. In the case of the metaverse, where most of these virtual realities are expected to be more emphasised, AR will allow cultural sites to attract more tourism interest, including those in remote and little-known places which seldom receive this kind of attention.

The construction industry is another critical sector in existing and emerging smart cities, especially in the pursuit of optimising available urban spaces, particularly in ensuring the growing urban population is housed, and factoring in dimensions like recreation, transportation and amenities (Cheng et al., 2021). This industry is also critical during the revitalisation of existing infrastructure and urban components to make them more competitive and applicable to changing urban frontiers. On this front, AR technology is argued to hold significant potential to enhance the industry, especially in allowing stakeholders to deploy the use of simulations of different aspects of planning, modelling, interior organisation and application of diverse ICT products (Kaji et al., 2018). In particular, using this technology, different players in the construction industry, including architects, local authorities, designers, builders and local communities, could have the opportunity to virtually interact with different models before they are actualised (Adăscăliţei & Bălţoi, 2018), hence ensuring that the resulting projects are universally supported and accepted.

6G-ENABLING TECHNOLOGIES AND THEIR ROLE IN FUTURE SMART CITIES

The sixth generation (6G) is expected to prompt a total paradigm shift to global wireless communication, especially in addressing the shortcomings of 5G and its predecessors. Already, as mentioned above, 5G has brought numerous improvements to 4G, and the results are evident in a wide scope of urban dimensions. However, it is still believed that 5G, despite its celebrated capabilities, does not effectively solve the issues of higher system capacity, low latency demand, quality of service and higher data rate that will be addressed by 6G (Chowdhury et al., 2019). In terms of speed, 6G promises at least 1,000 (terabyte) Gbps. The latency will be very low, approximately 100 microseconds, which is 50 times lower than the same in 5G. The technology further promises to be two times better in terms of energy efficiency, 10 times superior in reliability and two times better in spectral efficiency (Samsung, 2020).

Most importantly, the 6G technology is expected to inspire the emergence of even more advanced technologies with the capacity to improve experiences in smart cities. For instance, it is anticipated that it will influence the emergence of truly immersive extended reality, prompt the emergence of high-fidelity mobile holograms and influence the advancement of digital twins (Samsung, 2020). Other technologies that the proponents of 6G believe will experience unprecedented advancement include AI (which will advance to comprehensive AI), 3D printing, blockchain, cloud computing, quantum communication and others. In addition, it is projected that new technologies including spectrum sharing, split computing, terahertz communications, cell free communication, novel antenna technologies, high precision networks and others will emerge and become mainstay, especially in smart cities (Chowdhury et al., 2019).

In the previous chapters, the importance of data collection, storage, computation, analysis and sharing was covered at length. Nonetheless, it was highlighted how numerous challenges might continue to hamper the full exploitation of data due to capacity and storage. This notwithstanding, proponents of the 6G technology believe that most of those challenges will be effectively addressed by the emergence of technologies such as split computing which will allow for the simultaneous use of different computing technologies within a network instead of relying on homogenous computing technology. This will facilitate increased data-handling capacity, especially as by the time 6G is launched and market ready, the number of IoT devices and other data-generating products will have increased exponentially. It is estimated that the number of connected IoT devices will have reached a high of 125 billion by 2030, during the same period in which 6G is projected to enter the wireless communication market. Such an exponential increase will prompt a massive increase in data which will need to be analysed in real time to allow quick decisions to support automated urban dimensions, which experts predict will have increased significantly. For instance, it is projected that over 73 million people in the United States alone will lose their jobs due to automation (Davidson, 2017). In the United Kingdom it is projected that automation will lead to a 30 per cent loss of current jobs (Cameron, 2020). Globally, a report by McKinsey highlights that at least 375 million people will become victims of automation in different occupations (Manyika et al., 2017). For the record, most automation agendas are anticipated to be focused on sectors such as transport and health, where pilot tests, especially with regard to autonomous vehicles, have already taken place in diverse cities across the globe (Dixit et al., 2021). With the advancement of 5G, and later 6G, it is expected that some of the hurdles like faster data transfer that are required to fully exploit this option will have been overcome (Wang et al., 2020). Among the anticipated technologies that will benefit from 6G technology, especially in the quest for autonomous vehicles, is *visible light communication*, which is championed as the solution for growing demands for wireless connectivity. In autonomous vehicles, this is expected to enhance vehicle-to-vehicle communication (Tariq et al., 2020), supported by AI – particularly machine learning technologies.

However, while people might lose jobs from conventional sectors, it is projected that different technologies, including 5G and 6G, will prompt a paradigm shift on urban modelling where new dimensions like the metaverse will become a reality. This will prompt the opening of new employment frontiers, mainly in the virtual realm, and will benefit from technologies like comprehensive (distributed) AI, cryptocurrency, 3D, extended reality, digital replica and blockchain that are expected to be part of the pillars anchoring the metaverse and other future smart city ecosystems.

The different technologies that are expected to emerge as a result of advancements in wireless cellular communications from 5G to 6G will however come at a cost to future cities. Already, in the case of 5G, it is estimated that approximately \$1 trillion is required to build and install prerequisite infrastructures for the technology (Guilford, 2019). The costs come as a result of demands in terms of deployment, physical groundwork and associated technologies and equipment. The cost will also be influenced by the regulations and legal requirements of different governing bodies. Security concerns are expected to prominently inflate the cost of this technology. With 6G expected to be far more advanced than 5G, the costs of implementing it are expected to be even higher, though no substantial estimate has been released thus far. Most of the existing infrastructures supporting previous wireless communication technologies, including 5G, might not be relevant to 6G, hence will need to be replaced. Similarly, some of the existing devices might require an upgrade to incorporate additional features for 6G, while numerous others might become obsolete. Observing that a majority of existing smart devices are already in their billions, substantial financial resources will be required either to initiate upgrades or replace the existing infrastructure (Chowdhury et al., 2019).

THE ANTICIPATED FUTURE CITIES, THEIR PROSPECTS AND IMPACTS

The smart city planning models are based on the availability of data and subsequent data analysis and manipulation to influence and enhance different urban aspects. Even in anticipated future cities, data will remain the cog for those urban fabrics, especially following the increased number of technologies and innovations that are making it possible to effectively gather, store, compute, analyse and draw insights from the data generated by different urban components. Projections and anticipations on the nature of future cities is that there will be massive investment and adoption of advanced technologies such as AR, digital twins, cryptocurrency, blockchain and advanced AI that will emerge especially after the roll-out of 6G. These are expected to play a critical role in helping cities become resilient and adapt to the diverse challenges that already exist and those that are expected to arise in the future. These will come about by trends such as increasing urban population, rapid urbanisation, climate change, uncertainties arising from the COVID-19 pandemic and geopolitical tensions like the current Russian–Ukraine war.

Regarding climate change, the latest reports from bodies that heralded the COP 26 summit in Glasgow 2021 (UNFCCC, 2021b) highlighted that the world will continue to experience unprecedented climate-related incidences. These would be prompted by rising global temperatures that experts, including the scientific community, argue will exceed the 2 degrees Celsius pre-industrial levels against the objectives set in the 2015 Paris Agreement (UNFCCC, 2021a). The report showed that unless the global community increased their efforts and attention to decarbonisation, it will be inevitable that temperatures will rise by almost 3 degrees Celsius by 2040 (UNFCCC, 2021b). Such a rise will be injurious to many cities across the globe, with those in coastal regions, low-lying lands and small island developing states experiencing massive challenges (Allam, 2019b). Already there are numerous challenges being reported, including in areas that were originally thought to be relatively secure. However, regions like small island developing states and coastal cities have started to endure unprecedented vulnerabilities including rising sea levels, forced climate migration, the destruction of infrastructure and properties and the loss of natural resources (OECD, 2014).

Therefore, the above-mentioned technologies will have a significant influence on helping equip future cities with the capacity to not only accommodate the increasing population, but also to ensure that those cities are sustainable, liveable and have increased economic opportunities. In cities where existing economic opportunities are already experiencing extreme climate threats and might not be sustainable in the near future or will not manage to meet all of the revenue demands due for climate mitigation programmes, the adoption of technologies is expected to prompt the emergence of new economic dimensions. For instance, in the coastal cities that have often relied on trade and tourism as their economic mainstay, but are now under extreme pressure from rising sea levels, erratic and increased weather events like floods, shifting to a service industry powered by technologies could provide an economic lifeline (UNEP, 2014). This is not a new phenomenon, as it has been witnessed in cities like Singapore, which adopted a paradigm shift in its economy to become a service-oriented economy rather than the traditional product-based one (Allam & Allam, 2020). Following this shift, this island-city managed to improve its per capita income from less than $1,000 recorded before 1970 (it was $926 in 1970, which was also the highest before then) to over $65,641 in 2019. This is directly related to the service-oriented economy that the country adopted (Macro Trends, 2022).

Cities were the most affected during the COVID-19 pandemic. This is partly because they host most of the global population. Further, they are also the main economic engine of most economies, hence with trends like lockdowns and scaled-down economic activities, they could not comfortably sustain the residents therein nor those in the periphery (rural and semi-urban areas). The urgent actions taken on transportation, human–human interactions, working in offices, etc. greatly disrupted urban lifestyles, putting residents into disarray and prompting uncertainty (Sharifi & Khavarian-Garmsir, 2020). But, as argued in this chapter, cities that had some element of technology already deployed were seen to have higher levels of resilience that made it possible for various activities to continue, albeit at a considerably slower pace. Similar kinds of disruption cannot be ruled out in the future. However, with urban managers and other stakeholders increasing their attention to incorporate more smart elements and technologies in cities, the expectation is that some form of resilience and adaptability will be developed. On this, some of the notable technologically oriented approaches that are seen to be gaining traction post-pandemic include the increased attention on the 15-minute city concept that proposes that urban neighbourhoods should be based on travel time (chrono-urbanism) (Allam et al., 2022, forthcoming; Moreno, 2019). That is, making sure cities have all the basic components accessible by urban residents through cycling or walking in a time frame not exceeding 15 minutes (Moreno, 2019). This would have the result of eliminating the need for private cars, which not only contributes greatly to carbon dioxide emissions but also hinders robust human interaction. On this, with the availability of technologies, it is anticipated that cities will manage to give priority to human-scale objectives, thus even in times of pandemic, people will be better prepared than they were in 2020. Another notable model that will become part of urban fabrics is the metaverse, which is expected to make it possible for urban dwellers to seamlessly thrive in both the physical and virtual realms (Allam, forthcoming; Prathap & Gill, 2021; Theo, 2021), hence overcoming major urban social challenges like reduced human interaction and limited time for self-actualisation.

From an economic point of view, future cities might prove expensive in terms of the financial and resource investments that will be required to ensure that different technologies are sufficiently incorporated. Nevertheless, when this challenge is finally overcome, it might be possible to resolve climate change, socioeconomic challenges and environmental vulnerabilities that are synonymous with cities. After all, it is expected that most cities will increase their economic opportunities (both in the physical and virtual environments) to cater for increasing global unemployment rates, largely accelerated by the pandemic. However, such opportunities will be tied to how existing and emerging technologies will be utilised and intertwined together.

REFERENCES

Adăscăliței, I., & Bălțoi, I.-C.-M. (2018). The influence of augmented reality in construction and integration into smart city. *Informatica Economica, 22*(2), 55–67.

Ahmad, I., Kumar, T., Liyanage, M., Okwuibe, J., Ylianttila, M., & Gurtov, A. (2017, 18–20 September). 5G security: Analysis of threats and solutions. 2017 IEEE Conference on Standards for Communications and Networking.

Akhtar, S. (2005). 2G-4G Networks. In M. Pagani (Ed.), *Encyclopedia of Multimedia Technology and Networking* (pp. 964–973). IGI Global.

Allam, Z. (2019a). Achieving neuroplasticity in artificial neural networks through smart cities. *Smart Cities, 2*(2). https://doi.org/10.3390/smartcities2020009

Allam, Z. (2019b). *Cities and the Digital Revolution: Aligning Technology and Humanity.* Springer Nature.

Allam, Z. (2020a). The first 50 days of COVID-19: A detailed chronological timeline and extensive review of literature documenting the pandemic. In Z. Allam (Ed.), *Surveying the COVID-19 Pandemic and Its Implications* (pp. 1–7). Elsevier.

Allam, Z. (2020b). The second 50 days: A detailed chronological timeline and extensive review of literature documenting the COVID-19 pandemic from day 50 to day 100. In Z. Allam (Ed.), *Surveying the COVID-19 Pandemic and Its Implications* (pp. 9–39). Elsevier.

Allam, Z. (2020c). *The Rise of Autonomous Smart Cities: Technology, Economic Performance and Climate Resilience.* Springer International Publishing.

Allam, Z. (2020d). The third 50 days: A detailed chronological timeline and extensive review of literature documenting the COVID-19 pandemic from day 100 to day 150. In Z. Allam (Ed.), *Surveying the COVID-19 Pandemic and Its Implications* (pp. 41–69). Elsevier.

Allam, Z. (2021). On complexity, connectivity and autonomy in future cities. In *The Rise of Autonomous Smart Cities: Technology, Economic Performance and Climate Resilience* (pp. 31–47). Springer International Publishing.

Allam, Z., & Allam, Z. (2020). The rise of Singapore. In *Urban Governance and Smart City Planning* (pp. 1–26). Emerald Publishing.

Allam, Z., & Jones, D. S. (2021). Future (post-COVID) digital, smart and sustainable cities in the wake of 6G: Digital twins, immersive realities and new urban economies. *Land Use Policy, 101*, 105201.

Allam, Z., Bibri, S. E., Jones, D. S., Chabaud, D., & Moreno, C. (2022). Unpacking the '15-minute city' via 6G, IoT, and digital twins: Towards a new narrative for increasing urban efficiency, resilience, and sustainability. *Sensors, 22*(4). https://doi.org/10.3390/s22041369

Allam, Z., Ayyoob, S., Bibri, S. E., Jones, D. S., & Krogstie, J. (2022). The Metaverse as a Virtual Form of Smart Cities: Opportunities and Challenges for Environmental, Economic, and Social Sustainability in Urban Futures, *Smart Cities, 5*, 771–801.

Berger, E., Israel, G. L., & Miller, C. (2016). *World History: Cultures, States, and Societies to 1500.* University Press of North Georgia.

Britannica. (2020). Population change. Britiannica. Retrieved 10 March 2022 from www.britannica.com/topic/modernization/Population-change

Cameron, E. (2020). How will automation impact jobs? PWC. Retrieved 10 March 2020 from www.pwc.co.uk/services/economics/insights/the-impact-of-automation-on-jobs.html

Chapman, M. (2021, 2 November). Is the metaverse a friend or foe to travel? Phocus Wire. Retrieved 4 December 2021 from www.phocuswire.com/Metaverse-friend-foe-travel

Charr, M. (2020, 27 March). What can AR do to bring heritage sites to life? Museum Next. Retrieved 12 March 2022 from www.museumnext.com/article/ what-can-ar-do-to-bring-heritage-sites-to-life/

Cheng, H., Wang, F.-F., Dong, D.-W., Liang, J.-C., Zhao, C.-F., & Yan, B. (2021). Effects of smart city construction on the quality of public occupational health: Empirical evidence from Guangdong province, China. *Frontiers in Public Health*, *9*. www.frontiersin.org/article/10.3389/fpubh.2021.769687

Chowdhury, M. Z., Shahjalal, M., Ahmed, S., & Jang, Y. M. J. a. p. a. (2019). 6G wireless communication systems: Applications, requirements, technologies, challenges, and research directions. *IEEE*, *1*, 957–975.

Davidson, P. (2017, 1 December). Automation could kill 73 million US jobs by 2030. Transport Topics. Retrieved 10 March 2020 from www.ttnews.com/ articles/automation-could-kill-73-million-us-jobs-2030

Di Giulio, R., Boeri, A., Longo, D., Gianfrate, V., Boulanger, S. O. M., & Mariotti, C. (2019). ICTs for accessing, understanding and safeguarding cultural heritage: The experience of INCEPTION and ROCK H2020 projects. *International Journal of Architectural Heritage*, *15*(6), 1–19.

Dixit, A., Kumar Chidambaram, R., & Allam, Z. (2021). Safety and risk analysis of autonomous vehicles using computer vision and neural networks. *Vehicles*, *3*(3).

Facebook. (2021, 28 October). Connect 2021: Our vision for the metaverse. Meta. Retrieved 1 December 2021 from https://tech.fb.com/connect-2021-our -vision-for-the-metaverse/

Galazzo, R. (2020, 21 September). Timeline from 1G to 5G: A brief history on cell phones. CENGN. Retrieved 10 March 2022 from www.cengn.ca/information -centre/innovation/timeline-from-1g-to-5g-a-brief-history-on-cell-phones/

Globenewswire. (2022, 22 February). Global smart cities market to reach $2.5 trillion by 2026. Globe News Wire. Retrieved 11 March 2022 from www .globenewswire.com/news-release/2022/02/22/2389027/0/en/Global-Smart -Cities-Market-to-Reach-2-5-Trillion-by-2026.html

Guilford, G. (2019, 28 September). The coming 5G revolution. Quartz. Retrieved 10 March 2022 from https://qz.com/1716218/to-build-out-5g-mobile-wireless -operators-must-spend-up-to-1-trillion-without-much-in-the-way-of-return/

Hall, S. B., & Li, C. (2021, 29 October). What is the metaverse? And why should we care? World Economic Forum. Retrieved 2 December 2021 from www .weforum.org/agenda/2021/10/facebook-meta-what-is-the-metaverse/

International Telecommunication. (2011, 4 April). Evolution of the mobile market. International Telecommunication Union. Retrieved 10 March 2022 from www.itu.int/osg/spu/ni/3G/technology/

Kaji, S., Kolivand, H., Madani, R., Salehinia, M., & Shafaie, M. (2018). Augmented reality in smart cities: Applications and limitations. *Journal of Engineering Technology*, *6*(1), 18.

Kaneda, T., Greenbaum, C., & Kline, K. (2020). 2020 world population data sheet. Report. PRB. Retrieved 2 February 2022 from www.prb.org/wp -content/uploads/2020/07/letter-booklet-2020-world-population.pdf

Khan, A. H., Qadeer, M. A., Ansari, J. A., & Waheed, S. (2009, 3–5 April). 4G as a next generation wireless network. 2009 International Conference on Future Computer and Communication.

Macro Trends. (2022). Singapore GDP per capita 1960–2022. Macrotrends. Retrieved 12 March 2022 from www.macrotrends.net/countries/SGP/ singapore/gdp-per-capita

Manyika, J., Lund, S., Chui, M., Bughin, J., Woetzel, J., Batra, P., Ko, R., & Sanghv, S. (2017). Jobs lost, jobs gained: workforce transitions in a time of automation. McKinsey. Retrieved 2 February 2022 from www.mckinsey.com/~/media/ mckinsey/industries/public%20and%20social%20sector/our%20insights/what %20the%20future%20of%20work%20will%20mean%20for%20jobs%20skills %20and%20wages/mgi-jobs-lost-jobs-gained-executive-summary-december-6 -2017.pdf

Mascitelli, B., & Chung, M. (2019). Hue and cry over Huawei: Cold war tensions, security threats or anti-competitive behaviour? *Research in Globalization, 1,* 100002.

Moore, D. (2020, 19 May). How 5G will power smart cities of the future. *The Globe and Mail.* Retrieved 11 March 2022 from www.theglobeandmail.com/ featured-reports/article-how-5g-will-power-smart-cities-of-the-future/

Moreno, C. (2019, 30 June). The 15 minutes-city: for a new chrono-urbanism! Retrieved 10 November 2020 from www.moreno-web.net/the-15-minutes-city -for-a-new-chrono-urbanism-pr-carlos-moreno/

Nawaz, F., Ibrahim, J., Junaid, M., Kousar, S., Parveen, T., & Ali, M. A. (2020). A review of vision and challenges of 6G technology. *International Journal of Advanced Computer Science and Applications, 11*(2), 643–649.

Nayak, S., & Patgiri, R. (2020). 6G communication technology: A vision on intelligent healthcare. *arXiv.* Retrieved 2 February 2022 from https://arxiv.org/ abs/2005.07532

OECD. (2014). Cities and climate change: National governments enabling local action. In *OECD Policy Perspectives* (pp. 1–21). OECD.

Oughton, E. J., Frias, Z., van der Gaast, S., van der Berg, R. J. T., & Informatics. (2019). Assessing the capacity, coverage and cost of 5G infrastructure strategies: Analysis of the Netherlands. *Telematics and Informatics, 37,* 50–69.

Pallavicini, F., Chicchi Giglioli, I. A., Kim, G. J., Alcañiz, M., & Rizzo, A. (2021). Editorial: Virtual reality, augmented reality and video games for addressing the impact of COVID-19 on mental health. *Frontiers in Virtual Reality, 2.* www.frontiersin.org/article/10.3389/frvir.2021.719358

Prathap, M., & Gill, P. (2021, 1 October). The 'metaverse' may change the way you earn money, shop or even chill with friends in the future. *Business Insider.* Retrieved 1 December 2021 from www.businessinsider.in/investment/news/ the-metaverse-may-change-the-way-you-earn-money-shop-or-even-chill-with -friends-in-the-future/slidelist/86677906.cms

Rahman, A., Arabi, S., & Rab, R. (2021). Feasibility and challenges of 5G network deployment in least developed countries (LDC). *Wireless Sensor Network, 13,* 1–16.

Reach and Markets. (2021, 10 December). Global 5G services market report 2021–2025 and 2030 featuring AT&T, Verizon, China Mobile, Vodafone, Telstra, China Telecom, Deutsche Telecom, SK Telecom, Saudi Telecom, and T-Mobile. *Research and Markets.* Retrieved 11 March 2022 from www .globenewswire.com/news-release/2021/12/10/2350105/28124/en/Global-5G

-Services-Market-Report-2021-2025-2030-Featuring-AT-T-Verizon-China
-Mobile-Vodafone-Telstra-China-Telecom-Deutsche-Telecom-SK-Telecom
-Saudi-Telecom-and-T-Mobile.html

Ritchie, H., & Roser, M. (2019, September). Urbanization. Our World in Data. Retrieved 1 June 2021 from https://ourworldindata.org/urbanization#:~:text =Using%20these%20definitions%2C%20it%20reports,more%20than%206.1 %20billion%20people

Rong, B., Han, S., Kadoch, M., Chen, X., & Jara, A. (2020). Integration of 5G networks and Internet of Things for future smart city. *Wireless Communications and Mobile Computing*, Doi:10.1155/2020/2903525.

Sampaio, M., Navarro Haro, M. V., De Sousa, B., Vieira Melo, W., & Hoffman, H. G. (2021). Therapists make the switch to telepsychology to safely continue treating their patients during the COVID-19 pandemic. Virtual reality telepsychology may be next. *Frontiers, 1*. https://doi.org/10.3389/frvir.2020.576421

Samsung. (2020). *6G:* The next hyper-connected experience for all. Samsung Research. Retrieved 21 July 2020 from https://cdn.codeground.org/nsr/ downloads/researchareas/6G%20Vision.pdf

Sharifi, A., & Allam, Z. (2021). On the taxonomy of smart city indicators and their alignment with sustainability and resilience. *Environment and Planning B: Urban Analytics and City Science*, *49*(5), 1536–1555. https://doi.org/10 .1177/23998083211058798.

Sharifi, A., & Khavarian-Garmsir, A. R. (2020). The COVID-19 pandemic: Impacts on cities and major lessons for urban planning, design, and management. *Science of the Total Environment*, *749*, 142391.

Siani, A., & Marley, S. A. (2021). Impact of the recreational use of virtual reality on physical and mental wellbeing during the COVID-19 lockdown. *Health Technol (Berl)*, *11*(2), 425–435.

Statista Research Department. (2016, 27 November). Internet of Things – number of connected devices worldwide 2015–2025. Statista. Retrieved 13 December 2019 from www.statista.com/statistics/471264/iot-number-of-connected -devices-worldwide/

Tariq, F., Khandaker, M. R., Wong, K.-K., Imran, M. A., Bennis, M., & Debbah, M. J. I. W. C. (2020). A speculative study on 6G. *IEEE*, *27*(4), 118–125.

Theo. (2021, 5 August). Digital twins, IoT and the metaverse. *Medium*. Retrieved 3 December 2021 from https://medium.com/@theo/digital-twins-iot-and-the -metaverse-b4efbfc01112

UNEP. (2014). Emerging issues for small island developing states. UNEP. Retrieved 1 February 2022 from https://sustainabledevelopment.un.org/ content/documents/1693UNEP.pdf

UNESCO. (2017). Launch of the 2018 global report. Retrieved from https:// en.unesco.org/creativity/sites/creativity/files/global_report_fact_sheet_en.pdf

UNESCO. (2021). Creative cities network. UNESCO. Retrieved 17 September 2021 from https://en.unesco.org/creative-cities/home

UNFCCC. (2021a, 26 February). 'Climate commitments not on track to meet Paris Agreement goals' as NDC synthesis report is published. UNFCCC. Retrieved 24 September 2021 from https://unfccc.int/news/climate-commitments-not -on-track-to-meet-paris-agreement-goals-as-ndc-synthesis-report-is-published

UNFCCC. (2021b, 17 September). Nationally determined contributions under the Paris Agreement: Synthesis report by the secretariat. Conference of the Parties, meeting of the parties to the Paris Agreement, Glasgow. Retrieved 1 February 2022 from https://unfccc.int/sites/default/files/resource/cma2021 _08_adv_1.pdf

Wang, M., Zhu, T., Zhang, T., Zhang, J., Yu, S., & Zhou, W. (2020). Security and privacy in 6G networks: New areas and new challenges. *Digital Communications and Networks*, *6*(3), 281–291.

Ward, K. (2019, 18 February). Timeline of revolutions. Manufacturing Data Summit. Retrieved 10 March 2022 from https://manufacturingdata.io/ newsroom/timeline-of-revolutions/#:~:text=The%20Third%20Industrial %20Revolution%2C%20or,the%20discovery%20of%20nuclear%20energy

Yang, C., Liang, P., Fu, L., Cui, G., Huang, F., Teng, F., & Bangash, Y. A. (2022). Using 5G in smart cities: A systematic mapping study. *Intelligent Systems with Applications*, *14*, 200065.

7. The emergence of a new urban proximity-based morphology: the 15-minute city and the smart city

INTRODUCTION

In the past few decades, numerous urban planning models have emerged, all targeted at making existing and emerging cities more sustainable, liveable and centres for excellence in terms of economic growth, social pursuits and governance. The different models have been accelerated by various factors, including the growing urban population. The projection is that at least 2 billion more people will be added to the global population by 2050; prompting an increase from the current 7.7 billion to 9.7 billion people (Roser, 2019). Of these, 70 per cent will be living in cities across the globe (Kaneda et al., 2020). Urban planning models have further been influenced by global trends such as increased demand for different resources (energy, construction materials, water, food supply, manufactured products) that have triggered frequent unsustainable practices and outcomes. For instance, in the energy sector, the increasing demand has over the years prompted an increased overconsumption of non-renewable products like fossil fuels. Essentially, these have been identified as the leading cause for unprecedented greenhouse gas emissions (Allam, 2018; Covert et al., 2016; Gross, 2020). Globally, it is reported that 78–80 per cent of energy is generated from fossil fuels, which prompts over 60 per cent of all greenhouse gas emissions globally (United Nations, 2020).

As from 2015 when a score of ground-breaking global policy-oriented documents were launched, the smart city planning model became the most popular urban planning model (Allam & Newman, 2018; Batty, 2013; Lazaroiu & Roscia, 2012; Shelton et al., 2014). The creative city (Scott, 2006), sustainable city (Bibri & Krogstie, 2017, 2020) and sponge city models gained substantial traction, however, the smart city concept gained considerable traction, due to its accommodation with existing and emerging advanced technologies, which were also gaining in popularity during the same period (Allam, 2017; Sharifi & Allam, 2021). The number of connected Internet of Things (IoT) products and devices had grown to approximately 15.41 billion by 2015 and by 2020 the number had increased to approximately 42.62 billion devices (Statista Research Department, 2016). It is estimated that the positive growth of these devices, especially after the emergence of COVID-19, will spike them to a high of more than 75 billion products by 2025 (Stat Investor, 2022). Artificial intelligence (AI) technology, which has played an equally significant role in the growth of the smart city concept, has also been growing steadily over the years, reaching a market valuation of about $22.59 billion in 2020 from a low of $10.1 billion recorded in 2018. By 2025, this technology is expected to grow in its market value to over $126 billion (Statista Research Department, 2022). Similar positive growth has been reported with other smart technologies, which essentially have contributed to the immense growth of the smart city market. For instance, it is estimated that the smart city market was valued at approximately $517.6 billion in 2019. However, with an exponential compounded annual growth rate of 23 per cent, its value is projected to reach a high of $2,118.14 billion by 2024. Markets like Europe will contribute approximately 40 per cent of this growth (Technavio, 2022). Some of the contributing factors for this growth are reduced prices of connected IoT devices, increased adoption of renewable energy and reduction in prices of supportive technology such as AI and cloud computing. The popularity of this urban planning model could also be attributed to spirited efforts by different ICT corporations, both large and startups, in their substantial investments in terms of resources, research and expertise.

This model has influenced the realisation of many positive impacts on urban environments, especially in helping to reduce emissions through the adoption of products like smart meters and smart grids in the energy sector. It has also been supportive in making possible the adoption of trends like ride sharing (Tao et al., 2021), alternative mobility options like the use of bicycles (Pappalardo et al., 2017) and the use of electric vehicles (Soltani-Sobh et al., 2017) in the transport sector. However, it has also attracted criticism. One major concern is that though it has the potential to promote sustainable efforts, it has somehow contributed to the rise of various unsustainable practices, especially in the manufacturing of products and devices that support the model. Further, the model is also argued to fall short in its capacity to promote social dimensions, including promoting inclusivity, helping to reduce socioeconomic inequality and promoting healthy lifestyles (Cavada, 2022; Moreno et al., 2021; Özdemir et al., 2019). Some of these shortcomings became apparent during the height of COVID-19, when much of the global community especially in cities was put under lockdown (Troisi et al., 2022). Whereas some smart aspects like e-commerce, cashless transactions and virtual communication made it possible for people to have some relief, most social outcomes like interperson interaction and access to basic amenities were greatly hampered (Cavada, 2022). As a result, post-pandemic, there is a need for more consideration and a rethink on how the smart city concept can continue to be deployed, but with the social dimension being fully and carefully integrated as one of the core objectives of the model.

The pursuit to make the smart city concept embrace the social dimension can greatly benefit from exploring emerging planning concepts that focus on human-scale dimensions (Allam, 2020b). Among these is the 15-minute city concept, which was proposed in 2006 by Carlos Moreno but came into the limelight in 2019 after it was embraced in Paris (Moreno, 2016b; Moreno et al., 2021). This concept proposes the restructuring of cities to allow the placing of basic amenities within a radius where each of those could be accessed within a 15-minute time frame through walking or cycling (Moreno, 2019). As will be comprehensively addressed in the next section, this planning concept, which could be viewed as an offshoot of the smart city concept, is an interesting candidate that can advance the objectives already underpinned in the smart city planning model. The model is fashioned to achieve four key social objectives. These include proximity, density, diversity and digitalisation. A seamless combination of these four dimensions is championed to have the capacity to assist in solving the major urban social challenges that the COVID-19 pandemic made apparent. Although the concept is relatively new, it holds potential advantages in addressing not only the social dimension, but also helping to achieve the sustainability and resilience that previous planning models have not managed to address. In light of this background, this chapter will explore the technical dimensions that can potentially lead to the adoption and implementation of the 15-minute city.

THE 15-MINUTE CITY

The 15-minute city is a relatively new planning concept that was conceived and advanced by Carlos Moreno in 2006 (Allam et al., 2020; Moreno et al., 2021). The concept champions the rethinking of urban morphologies to allow a planning departure from the conventional philosophy of placing different urban elements in relation to how they affect and impact vehicular movement (Allam et al., 2022b). The underlying theme in planning according to the 15-minute city model should be human pursuits, where social aspects such as human interaction, accessibility to basic amenities, health and self-actualisation are given maximum attention. The ultimate objective is to deliberately plan and position different urban nodes, infrastructures and opportunities in a way that they can be reached and utilised by urban residents within a time frame not exceeding 15 minutes. This then discourages the need for private cars or fossil fuel-dependent mobility options for short commutes, hence limiting emissions that compromise sustainability agendas. The 15-minute city concept further seeks to ensure that urban residents have the maximum time and opportunities to create cordial social relationships (Allam et al., 2022b; C40 Cities Climate Leadership Group, 2020). That is, by helping reduce the need for car travel as well as making it possible to access almost all amenities in a short commute, urban residents will have the opportunity to interact as they walk, cycle or stroll through parks. Further, the amount of time saved becomes an important aspect in ensuring that created relationships are not casual but cordial. This can only be achieved if people have maximum time to interact and socialise.

The 15-minute city model is based on the chrono-urbanism concept where the aspect of time is accorded maximum attention relative to space (Moreno, 2016a). This explains why proponents of the concept emphasise the need for a 15-minute radius for an urban neighbourhood, even with clear knowledge that some cities are naturally endowed with maximum space. By overcoming the aspect of space as the motivating factor in placing different urban nodes, it is then possible to pursue the four distinct characteristics that Moreno et al. (2021) perceive to be critical for cities to uphold the aspect of liveability and, by extension, manage to accelerate the achievement of Sustainable Development Goal (SDG) 11 (United Nations Department of Economic and Social Affairs, 2021), and also the precepts of the New Urban Agenda (United Nations, 2017). These characteristics (pillars) include diversity, proximity, density and ubiquitousness. Further, the aspect of digitalisation is viewed as being another characteristic of the model.

With regard to the aspect of urban density, the 15-minute city champions the need for current and future cities to be made up of neighbourhoods that have a distinct number of residents (Moreno et al., 2021). That is, within a 15-minute radius neighbourhood, a pre-determined number of residents or households are facilitated in terms of housing, amenities and security such that there is no overutilisation and/or underutilisation of available resources. This is critical as it would allow for the optimal use of available resources and informed provision of 'basic' urban needs in a way that the city management would not be strained. Thus, it might become possible to address cases of informal neighbourhoods, urban sprawl, homelessness and other social challenges. Such would further be addressed by ensuring that the 15-minute city concept is made ubiquitous – readily available both in terms of quantity and quality. This factor is also critical in addressing one of the downsides of the smart city concept where it has been found to catalyse unprecedented urban revitalisation, which eventually leads to gentrification (Sisson, 2020). In the course of restructuring cities to accommodate the principles of the 15-minute city concept, there will be massive urban revitalisation and conservation while ensuring those projects address gentrification concerns.

The diversity dimension propagated and advanced in the 15-minute city concept envisions urban neighbourhoods that are rich in cultural vibrancy, both in terms of urban structures as well as people drawn from different cultural backgrounds (Moreno, 2016a, 2020a). With respect to the urban demographic composition, proponents of this planning model state that cities are fashioned to allow cohesive co-existence between people regardless of their background, such as race, occupation, religion and culture (Moreno et al., 2021). Thus, it would essentially be possible to advance the aspect of inclusivity proposed in SDG 11. Further, the idea of rich cultural diversity would promote other positive urban qualities like harmonious living, promote economic opportunities supported by factors like cultural heritage and creative arts that are only possible in cases where diversity is promoted (Ager & Brückner, 2013). When focusing on diversity in urban structures, the anticipation is to have urban infrastructures and other elements that could be used for multiple purposes, allowing for sustainability, the optimal use of scarce urban spaces and further reducing the need for travel. In the book *The Death and Life of Great American Cities* by Jane Jacobs (1961), one of her concerns was how conventional planning models were leading to the destruction of neighbourhoods by giving too much attention to modernist planning approaches, which demonstrate little concern for the social dimension. The ultimate consequences were the emergence of disjointed cities that disrupt the lifestyles of locals. The diversity dimension championed in the 15-minute model seeks to rewrite those ills by ensuring that the plight of locals, and their demands, are factored into the design and implementation of diverse projects. Therefore, the 15-minute city concept has the potential to promote public participation (C40 Cities Climate Leadership Group, 2020).

After the emergence of COVID-19 and the subsequent containment measures that were introduced by different governments across the globe (Allam, 2020c, 2020d, 2020e), it became evident that most urban residents were accustomed to using automobiles to reach different urban nodes. The obsession with vehicles was expressed in Marchetti's Constant (Newman & Jennings, 2012) as making people 'tend to travel further'. But post-pandemic and into the future, the travelling obsession is expected to become unfashionable as many activities like work, shopping, recreation and entertainment are expected to happen virtually. Already, some models like working from home (most popular during the height of COVID-19), remote working and virtual communication have already gained substantial popularity since the outbreak of the pandemic.

SMART CITY NETWORKS IN THE '15-MINUTE CITY' CONCEPT

The smart city concept relies heavily on different technologies to address diverse urban challenges and to achieve the objectives of increasing liveability and resilience in cities (Allam, 2020a, 2020f; Biyik et al., 2021). The different technologies are linked together via complex networks that allow for the collection, storage, analysis and transfer of data obtained from different connected devices and sensors. These technologies are critical in facilitating the achievement of the different dimensions that form the basis for the concept of the 15-minute city (Woods, 2020). For instance, with respect to proximity, the wide array of connected devices and sensors are argued to have the capacity to allow the mapping of distances between different urban nodes, such as infrastructures and built environments where basic services are offered. This would facilitate for informed decision making on how different nodes could be fashioned to make them coherent with the theme of the 15-minute city. A case in point is the adoption of digital twins technology, which is gradually gaining traction in urban planning platforms. This technology has the capacity to facilitate digital representation of different physical urban objects in the virtual environment. Consequently, that would make it possible to simulate scenarios of how different planning options would impact the morphology and functioning of cities (Allam & Jones, 2021; Charitonidou, 2022). By so doing, the city would not only benefit from the most optimal option obtained after the simulation, but the cost of implementing such a morphological change would be relatively cheap compared to the past when such programmes were initiated in what can be termed as 'trial and error' (Deren et al., 2021). Further, the availability of models of a particular neighbourhood in the virtual realm would provide opportunities for other players in urban planning realms to interact with the digital replica, hence providing input on how they would like diverse projects to be implemented.

The availability of technologies such as IoT networks and devices, AI and cloud computing is expected to play a significant role in the social dimension. In particular, these will be critical in improving the security component of the cities, more so in social meeting places like parks, recreation spaces and entertainment hubs where urban residents are likely to meet (Martin, 2017). Security will further be paramount in walking paths and cycling lanes, especially following the recent increase in cases of fatalities and accidents that pedestrians have experienced in urban areas compared to rural areas. To illustrate this, it is reported that in 2019 alone, urban areas in America recorded approximately 54 per cent of traffic fatalities compared to 45 per cent reported in rural areas (National Highway Traffic Safety Administration, 2021). Such discrepancies are unsurprising as most of the global population, especially in developed countries like the United States, reside in urban areas (World Bank, 2020).

Technology further enhances security by allowing for the real-time sharing of information and data between law enforcement agencies, private security systems and the general public. After sharing, technology further allows for real-time analysis and dissemination of drawn conclusions and insights to relevant departments and agencies for quicker action. Besides security, smart city networks are also critical in the pursuit of the 15-minute city concept, especially in helping identify priority areas that urgently need to be changed. A case in point is Paris, where the local government identified the need to reduce the number of private vehicles in the city as a priority toward achieving the 15-minute city concept (Reid, 2020a, 2020b). Instead of creating additional car lanes, the government prioritised the creation of cycling infrastructure, hence ensuring that people could continue commuting despite the necessary disruption warranted by the removal of private cars (Campbell, 2019). As a result, Paris now has approximately 1,400 km of dedicated cycle routes and it is anticipated that an additional 180 km of new bike lanes will be ready by 2026 (Open Access Government, 2021), hence building the city's capacity to become 100 per cent 'cyclable' (O'Sullivan & Bliss, 2020).

The reality of COVID-19 and the subsequent disruption it caused further expresses the importance of technology in the 15-minute city setup (OECD, 2020; Petropoulos, 2021). For instance, during the height of the pandemic, it became untenable for a substantial number of the global population to continue working in their normal work environment (Kumar et al., 2021). Some reprieve was on those who could successfully transition to working from home (remotely). Unfortunately, for a majority of the global population, the pandemic prompted a total disruption on the work cycle (Shang et al., 2021), thus prompting increased cases of job loss and unemployment. Diverse technologies further facilitated a seamless transition in workplaces, the education system, entertainment, shopping and the financial sector during the height of the pandemic, thus preventing a total shutdown. The same benefits that technology offered to urban areas during COVID-19 could be escalated in the 15-minute city environments to allow for the achievement of specified objectives (Allam & Jones, 2021). For instance, by investing and deploying technologies such as AI, big data, augmented reality and digital twins make it possible for remote working, hence reducing the need for travel. This will have a positive impact on the agenda to reduce private cars in cities (Lachapelle et al., 2017).

The attainment of the 15-minute city concept is expected to be an expensive undertaking as it will require an almost total overhaul of urban morphologies. This is particularly true in ensuring that the proximity dimension is attained. The cost tag has some bearing on achieving the ubiquitousness dimension, which in turn helps in addressing the inclusivity agenda. However, by utilising the already existing smart city networks, the 15-minute city concept might benefit in helping create new job opportunities that could be performed remotely, hence making it possible for people living outside the 15-minute cycling or walking distance to work virtually. The ability to work remotely will make it possible to have an increased number of 15-minute neighbourhoods within a given city, all connected to each other via different smart technologies. This is paramount especially in preventing the negative impacts of gentrification that are associated with the 15-minute city concept, and which might disadvantage a significant number of urban residents who might not have the capacity to cope with lifestyles occasioned in the 15-minute city setup.

Going into the future, where there is anticipation of increased use of existing and emerging technologies such as immersive reality, 6G, digital twins, comprehensive AI, cryptocurrency and blockchain technologies, it is paramount for the 15-minute city to be designed in a way that could effectively utilise these technologies.

THE POSSIBLE CHALLENGES AND PROSPECTIVE SOLUTIONS TO IMPLEMENTING THE '15-MINUTE CITY' CONCEPT

The implementation of the 15-minute city concept is gaining considerable attention in different major cities, especially after the impacts of COVID-19. However, just like other urban planning models, its implementation is experiencing, and is expected to continue (at least in the short term) to experience, some hurdles (Deloitte, 2022). Examples of these challenges include financial constraints, insufficient public policies and frameworks to guide its implementation, apprehension from some stakeholders (especially local residents) and technicalities with different technologies.

On the financial constraints, the 15-minute city is not an isolated concept, but a continuation of what cities have been experiencing in their quest to implement different projects. What makes the 15-minute city model different, however is the magnitude of work that will need to be undertaken especially when the project is to be implemented in a city with already established infrastructures and a built environment. The 15-minute city model proposes an overhaul to allow for the creation of time-scaled neighbourhoods. This means that a city might have a number of different 15-minute city 'units', each with different basic amenities intentionally positioned to facilitate accessibility within the limited time span. Achievement of this feat thereby demands restructuring existing urban design thinking, and this is expected to attract huge financial resources and capital requirements. For instance, to create the extra 180 km of cycle lanes, the local government in Paris allocated approximately 250 million euros (The Local, 2021). Approximately 12.5 million euros had already been used in the rehabilitation of 70 housing apartments in one region within the city (the Minimes Barracks former parking lot) and more is expected to be consumed in other programmes including during the enforcement of the 50 per cent reduction in private cars in the city (Reid, 2020a). In a report published by Deloitte in 2013, it was rightly argued that a majority of urban areas have no financial capacity to undertake huge investment programmes from public budgetary allocations but instead have to rely on external sources and the private sector.

After the COVID-19 pandemic, this situation might have become even more complicated as most economies were greatly impacted, plunging some economies into a debt crisis (Sharifi & Khavarian-Garmsir, 2020; UNCTAD, 2021). This then complicates their capacity to effectively engage in any project that is capital intensive as is the case with the 15-minute city concept.

The paradigm shift from a conventional city planning model to the 15-minute city concept is an ambitious undertaking that requires sound policies and legal frameworks formulated after a wide-scope consultation with all relevant stakeholders. This argument is supported by criticism raised in different quarters, prompting notable setbacks in the implementation of some elements of such projects in a number of cities. For instance, in Paris, local government policy to drastically reduce the number of private cars and motorcycles from the city centre by 2024 faced public apprehension that prompted its delay for two years (The Local, 2021). The postponement of this programme was meant to allow more time for the enforcement to be furnished in a way that does not attract more negative attention than intended. However, an article published by the World Economic Forum (Ratti & Florida, 2021) helps express and uncover a variety of the reasons that could be associated with the public uproar on the implementation of the programme – not only in Paris but also in other cities in which the concept has been embraced. In Paris, counterarguments were that the removal of vehicles would prompt traffic congestion in outer areas of the city and also would deny residents living in suburbs viable public transport options (vehicles) that they need to get to work (The Local, 2021). This point is valid, as there are some unique urban infrastructures and institutions that cannot be replicated in each of the 15-minute neighbourhoods, for instance, universities, museums and some cultural heritage sites that are unique to a city. According to critics of the 15-minute city concept, some of those aforementioned institutions/urban assets are highly associated with cities, and relaxing the dynamism around them to allow for their replication at neighbourhood scale would water down their value and importance.

Whereas it is possible for urban residents to live without some of the aforementioned urban features and still accomplish the six functions (commerce, working, living, healthcare, entertainment and education) (Moreno, 2020b) identified as essential in a 15-minute city concept, it is without doubt that a sizeable number of urban residents earn their living working in them. Therefore, public policies focusing on promoting the 15-minute concept need to be robust such that the negative reactions do not overshadow the actions of its implementation (Allam et al., 2022a, 2022c). Although it is not possible to replicate some features of conventional cities in every single 15-minute neighbourhood, it is possible to ensure that residents of a neighbourhood have unmetered access to them when needed. This could be achieved by ensuring robust smart city technologies such as immersive reality, AI, digital twins and high-speed connectivity (e.g. 5G) are incorporated into the 15-minute city design. This would allow residents to virtually access different physical objects, and when a physical presence is required, alternative mobility options should be available that link different areas of a city. Such challenges can further be overcome by providing for active participation of local residents – incorporating their input and demands. Through public engagement, deliberate civic education can explain the importance and advantages of restructuring the city into different time-scaled units (Arnstein, 1969; Raco, 2009). This would not only impact the acceptability of the project but would also facilitate ownership of the project by residents. Civic education further needs to be accompanied by deliberate cooperation between different stakeholders in the urban planning realm to reduce friction between different groups. This would be particularly important in exploring alternative financial options that could be adopted to actualise the project.

REFERENCES

Ager, P., & Brückner, M. (2013). Cultural diversity and economic growth: Evidence from the US during the age of mass migration. *European Economic Review*, *64*, 76–97.

Allam, M. Z. (2018). *Redefining the Smart City: Culture, Metabolism and Governance. Case Study of Port Louis, Mauritius*. Curtin University.

Allam, Z. (2017). Building a conceptual framework for smarting an existing city in Mauritius. *Journal of Biourbanism*, *17*(1–2), 103–121.

Allam, Z. (2020a). Actualizing big data through revised data protocols to render more accurate infectious disease monitoring and modeling. *Surveying the Covid-19 Pandemic and Its Implications*, *71*.

Allam, Z. (2020b). Seeking liveability through the Singapore model. *Urban Governance Smart City Planning*. Emerald Publishing.

Allam, Z. (2020c). The first 50 days of COVID-19: A detailed chronological timeline and extensive review of literature documenting the pandemic. *Surveying the COVID-19 Pandemic and Its Implications*, *1*, 1–7.

Allam, Z. (2020d). The second 50 days: A detailed chronological timeline and extensive review of literature documenting the COVID-19 pandemic from day 50 to day 100. *Surveying the COVID-19 Pandemic and Its Implications*, *9*.

Allam, Z. (2020e). The third 50 days: A detailed chronological timeline and extensive review of literature documenting the COVID-19 pandemic from day 100 to day 150. *Surveying the COVID-19 Pandemic and Its Implications, 41*, 41–69.

Allam, Z. (2020f). Underlining the role of data science and technology in supporting supply chains, political stability and health networks during pandemics. *Surveying the COVID-19 Pandemic and Its Implications, 129*.

Allam, Z., & Jones, D. S. (2021). Future (post-COVID) digital, smart and sustainable cities in the wake of 6G: Digital twins, immersive realities and new urban economies. *Land Use Policy, 101*, 105201.

Allam, Z., & Newman, P. (2018). Redefining the smart city: Culture, metabolism and governance. *Smart Cities, 1*(1). https://doi.org/10.3390/smartcities1010002

Allam, Z., Moreno, C., Chabaud, D., & Pratlong, F. (2020). Proximity-based planning and the '15-minute city': A sustainable model for the city of the future. In *The Palgrave Handbook of Global Sustainability* (pp. 1–20). Springer International Publishing.

Allam, Z., Moreno, C., Chabaud, D., & Pratlong, F. (2022a). Proximity-based planning and the '15-minute city': A sustainable model for the city of the future. In *The Palgrave Handbook of Global Sustainability*. Palgrave.

Allam, Z., Nieuwenhuijsen, M., Chabaud, D., & Moreno, C. (2022b). The 15-minute city offers a new framework for sustainability, liveability, and health. *The Lancet Planetary Health, 6*(3), e181–e183.

Allam, Z., Bibri, S. E., Jones, D. S., Chabaud, D., & Moreno, C. J. S. (2022c). Unpacking the '15-minute city' via 6G, IoT, and digital twins: Towards a new narrative for increasing urban efficiency, resilience, and sustainability. *Sustainability, 22*(4), 1369.

Arnstein, S. R. (1969). A ladder of citizen participation. *Journal of the American Institute of Planners, 35*(4), 216–224.

Batty, M. (2013). Big data, smart cities and city planning. *Dialogues in Human Geography, 3*(3), 274–279.

Bibri, S. E., & Krogstie, J. (2017). Smart sustainable cities of the future: An extensive interdisciplinary literature review. *Sustainable Cities and Society, 31*, 183–212.

Bibri, S. E., & Krogstie, J. (2020). Environmentally data-driven smart sustainable cities: Applied innovative solutions for energy efficiency, pollution reduction, and urban metabolism. *Energy Informatics, 3*(1), 29.

Biyik, C., Allam, Z., Pieri, G., Moroni, D., O'fraifer, M., O'Connell, E., Olariu, S., & Khalid, M. (2021). Smart parking systems: Reviewing the literature, architecture and ways forward. *Smart Cities, 4*(2), 623–642.

C40 Cities Climate Leadership Group. (2020, July). How to build back better with a 15-minute city. C40. Retrieved 10 November 2020 from www .c40knowledgehub.org/s/article/How-to-build-back-better-with-a-15-minute -city?language=en_US#:~:text=In%20a%2015%2Dminute%20city'%2C %20all%20citizens%20are,and%20sustainable%20way%20of%20life

Campbell, H. (2019, 12 September). How Paris is actually walking the climate change walk. *Time*. Retrieved 10 November 2020 from https://time.com/ 5669067/paris-green-city/

Cavada, M. (2022). Evaluate space after COVID-19: Smart city strategies for gamification. *International Journal of Human–Computer Interaction*, 1–12. https://doi.org/10.1080/10447318.2021.2012383

Charitonidou, M. (2022). Urban scale digital twins in data-driven society: Challenging digital universalism in urban planning decision-making. *International Journal of Architectural Computing*, *20*(2), 238–253, doi: 10.1177/14780771211070005.

Covert, T., Greenstone, M., & Knittel, C. R. (2016). Will we ever stop using fossil fuels? *Journal of Economic Perspectives*, *30*(1), 117–137.

Deloitte. (2013). Funding options: Alternative financing for infrastructure development. Deloitte. Retrieved 2 February 2022 from www2.deloitte.com/content/dam/Deloitte/au/Documents/public-sector/deloitte-au-ps-funding-options-alternative-financing-infrastructure-development-170914.pdf

Deloitte. (2022). 15-minute city. Deloitte. Retrieved 20 March 2022 from www2.deloitte.com/global/en/pages/public-sector/articles/urban-future-with-a-purpose/15-minute-city.html

Deren, L., Wenbo, Y., & Zhenfeng, S. (2021). Smart city based on digital twins. *Computational Urban Science*, *1*(1), 4.

Gross, S. (2020, June). Why are fossil fuels so hard to quit? Brookings Education. Retrieved 14 August 2021 from www.brookings.edu/essay/why-are-fossil-fuels-so-hard-to-quit/

Jacobs, J. (1961). *The Death and Life of Great American Cities*. Vintage Books.

Kaneda, T., Greenbaum, C., & Kline, K. (2020). 2020 world population data sheet. Report. PRB. Retrieved 2 February 2022 from www.prb.org/wp-content/uploads/2020/07/letter-booklet-2020-world-population.pdf

Kumar, P., Kumar, N., Aggarwal, P., & Yeap, J. A. L. (2021). Working in lockdown: The relationship between COVID-19 induced work stressors, job performance, distress, and life satisfaction. *Current Psychology*. https://doi.org/10.1007/s12144-021-01567-0

Lachapelle, U., Tanguay, G. A., & Neumark-Gaudet, L. (2017). Telecommuting and sustainable travel: Reduction of overall travel time, increases in non-motorised travel and congestion relief? *Urban Studies*, *55*(10), 2226–2244.

Lazaroiu, G. C., & Roscia, M. (2012). Definition methodology for the smart cities model. *Energy*, *47*(1), 326–332.

Martin, S. (2017). Real and potential influences of information technology on outdoor recreation and wilderness experiences and management. *Journal of Park and Recreation Administration*, *35*(1), 98–101.

Moreno, C. (2016a). La ville du quart d'heure: pour un nouveau chrono-urbanisme. *La tribune*. Retrieved 3 December 2020 from www.latribune.fr/regions/smart-cities/la-tribune-de-carlos-moreno/la-ville-du-quart-d-heure-pour-un-nouveau-chrono-urbanisme-604358.html

Moreno, C. (2016b, 5 October). The quarter-hour city: for a new chrono-urbanism. *La Tribune*. Retrieved 2 November 2021 from www.latribune.fr/regions/smart-cities/la-tribune-de-carlos-moreno/la-ville-du-quart-d-heure-pour-un-nouveau-chrono-urbanisme-604358.html

Moreno, C. (2019, 30 June). The 15 minutes-city: for a new chrono-urbanism! Retrieved 10 November 2020 from www.moreno-web.net/the-15-minutes-city -for-a-new-chrono-urbanism-pr-carlos-moreno/

Moreno, C. (2020a). *Droit de cité*. Humensis.

Moreno, C. (2020b, February). Urban proximity and the love for places chrono-urbanism, chronotopia, topophilia. Retrieved 5 January 2021 from www .moreno-web.net/urban-proximity-and-the-love-for-places-chrono-urbanism -chronotopia-topophilia-by-carlos-moreno/

Moreno, C., Allam, Z., Chabaud, D., Gall, C., & Pratlong, F. (2021). Introducing the '15-minute city': Sustainability, resilience and place identity in future post-pandemic cities. *Smart Cities*, *4*(1), 93–111.

National Highway Traffic Safety Administration. (2021, November). Rural/urban comparison of motor vehicle traffic fatalities. NHTSA. Retrieved 10 March 2022 from https://crashstats.nhtsa.dot.gov/Api/Public/ViewPublication/ 813206

Newman, P., & Jennings, I. (2012). *Cities as Sustainable Ecosystems: Principles and Practices*. Island Press.

O'Sullivan, F., & Bliss, L. (2020, 12 November). The 15-minute city: No cars required is urban planning's new utopia. Bloomberg. Retrieved 10 March 2022 from www.bloomberg.com/news/features/2020-11-12/paris-s-15-minute-city -could-be-coming-to-an-urban-area-near-you

OECD. (2020, 5 June). Building back better: A sustainable, resilient recovery after COVID-19. OECD. Retrieved 24 September 2021 from www.oecd .org/coronavirus/policy-responses/building-back-better-a-sustainable-resilient -recovery-after-covid-19-52b869f5/

Open Access Government. (2021, 13 October). How the 15-minute city concept can revolutionise London. Open Access Government. Retrieved 20 March 2022 from www.openaccessgovernment.org/how-the-15-minute-city-concept -can-revolutionise-london/122249/

Özdemir, A., Kourtit, K., & Nijkamp, P. (2019). Social policy in smart cities: The forgotten dimension. In N. Komninos & C. Kakderi (Eds), *Smart Cities in the Post-Algorithmic Era* (pp. 235–261). Edward Elgar Publishing.

Pappalardo, G., Stamatiadis, N., & Cafiso, S. (2017). Use of technology to improve bicycle mobility in smart cities. *IEEE*. https://doi.org/10.1109/ MTITS.2017.8005636

Petropoulos, G. (2021). Automation, COVID-19, and labor markets. ADBI Working Paper Series. ADB Institute. Retrieved 2 February 2022 from www .adb.org/sites/default/files/publication/688896/adbi-wp1229.pdf

Raco, M. (2009). Governance, urban. In R. Kitchin & N. Thrift (Eds), *International Encyclopedia of Human Geography* (pp. 622–627). Elsevier.

Ratti, C., & Florida, R. (2021, 11 November). The 15-minute city meets human needs but leaves desires wanting. Here's why. World Economic Forum. Retrieved 20 March 2022 from www.weforum.org/agenda/2021/11/15minute -city-falls-short/

Reid, C. (2020a, 28 June). Anne Hidalgo reelected as mayor of Paris vowing to remove cars and boost bicycling and walking. Forbes. Retrieved 5 November 2020 from www.forbes.com/sites/carltonreid/2020/06/28/anne-hidalgo

-reelected-as-mayor-of-paris-vowing-to-remove-cars-and-boost-bicycling-and
-walking/?sh=ba645d11c852

Reid, C. (2020b, 20 October). Anne Hidalgo to make good on pledge to remove half of city's car parking spaces. Forbes. Retrieved 10 November 2020 from www.forbes.com/sites/carltonreid/2020/10/20/paris-mayor-anne-hidalgo-to
-make-good-on-pledge-to-remove-half-of-citys-car-parking-spaces/?sh=
423352b916ec

Roser, M. (2019). Future population growth. Our World in Data. Retrieved 21 February 2021 from https://ourworldindata.org/future-population-growth#:~:
text=Population%20growth%20by%20world%20region,-More%20than%208
&text=The%20United%20Nations%20projects%20that,at%2010.9%20billion
%20by%202100

Scott, A. J. (2006). Creative cities: Conceptual issues and policy questions *Journal of Urban Affairs, 28*(1), 1–17.

Shang, Y., Li, H., & Zhang, R. (2021). Effects of pandemic outbreak on economies: Evidence from business history context. *Frontiers in Public Health, 9.* www.frontiersin.org/article/10.3389/fpubh.2021.632043

Sharifi, A., & Allam, Z. (2021). On the taxonomy of smart city indicators and their alignment with sustainability and resilience. *Environment Planning B: Urban Analytics City Science.* https://doi.org/10.1177/23998083211058798

Sharifi, A., & Khavarian-Garmsir, A. R. (2020). The COVID-19 pandemic: Impacts on cities and major lessons for urban planning, design, and management. *Science of the Total Environment, 749,* 142391.

Shelton, T., Zook, M., & Wiig, A. (2014). The 'actually existing smart city'. *Cambridge Journal of Regions, Economy and Society, 8*(1), 13–25.

Sisson, P. (2020, 21 September). What is a 15-minute city? City Monitor. Retrieved 10 November 2020 from https://citymonitor.ai/environment/what
-is-a-15-minute-city

Soltani-Sobh, A., Heaslip, K., Stevanovic, A., Bosworth, R., & Radivojevic, D. (2017). Analysis of the electric vehicles adoption over the United States. *Transportation Research Procedia, 22,* 203–212.

Stat Investor. (2022). Internet of Things – number of connected devices worldwide 2015–2025. Stat Investor. Retrieved 8 March 2022 from https://
statinvestor.com/data/33967/iot-number-of-connected-devices-worldwide/

Statista Research Department. (2016, 27 November). Internet of Things – number of connected devices worldwide 2015–2025. Statista. Retrieved 3 February 2021 from www.statista.com/statistics/471264/iot-number-of-connected-devices
-worldwide/

Statista Research Department. (2022, 17 March). Artificial intelligence software market revenue worldwide 2018–2025. Statista. Retrieved 8 March 2022 from www.statista.com/statistics/607716/worldwide-artificial-intelligence-market
-revenues/

Tao, Z., Nie, Q., & Zhang, W. (2021). Research on travel behavior with car sharing under smart city conditions. *Journal of Advanced Transportation, 1,* 1–13, doi: 10.1155/2021/8879908.

Technavio. (2022). Smart city market to grow at a CAGR of 23% by 2024. Cision. Retrieved 8 March 2022 from www.prnewswire.com/news-releases/

smart-city-market-to-grow-at-a-cagr-of-23-by-2024-increase-in-it-consolidation-
-modernization-to-boost-the-market-growth-17000-technavio-reports-301407606
.html

The Local. (2021, 24 October). How Paris will spend €250 million on making
city '100 % bike friendly'. The Local. Retrieved 21 March 2022 from www
.thelocal.fr/20211021/paris-e250-million-plan-to-make-city-100-cycleable/

Troisi, O., Fenza, G., Grimaldi, M., & Loia, F. (2022). COVID-19 sentiments in
smart cities: The role of technology anxiety before and during the pandemic.
Computers in Human Behavior, 126, 106986.

UNCTAD. (2021). Developing country external debt: From growing sustainability concerns to potential crisis in the time of COVID-19. UNCTAD. Retrieved
2 November 2021 from https://sdgpulse.unctad.org/debt-sustainability/

United Nations. (2017). New Urban Agenda. H. I. Secretariat. Retrieved 1
February 2022 from https://uploads.habitat3.org/hb3/NUA-English.pdf

United Nations. (2020). Cities and pollution. UN. Retrieved 1 November 2021
from www.un.org/en/climatechange/climate-solutions/cities-pollution

United Nations Department of Economic and Social Affairs. (2021). Goal 11:
Make cities and human settlements inclusive, safe, resilient and sustainable.
UNDESA. Retrieved 4 February 2022 from https://sdgs.un.org/goals/goal11

Woods, E. (2020, 4 August). The 15-minute city can also be a smart city. Guidehouse
Insights. Retrieved 15 December 2021 from https://guidehouseinsights.com/
news-and-views/the-15-minute-city-can-also-be-a-smart-city

World Bank. (2020, 20 April). Urban development: Overview. World
Bank. Retrieved 27 September 2021 from www.worldbank.org/en/topic/
urbandevelopment/overview#:~:text=With%20more%20than%2080%25
%20of,and%20new%20ideas%20to%20emerge

8. Future smart and autonomous cities: an overview toward future trends

INTRODUCTION

Since ancient times, human beings have significantly contributed to the development of cities, socially, economically and politically. Cities which emerged as early as 4,000 years ago in Indus Valley (present-day Pakistan and India) played key roles in the development of political, cultural and religious establishments that defined human social and economic power (Alex, 2020). After the emergence of the Industrial Revolution in the eighteenth century, the contribution of cities to different human endeavours became even more apparent, such that they still remain an important pillar in human civilisation. In terms of economic development, cities across the world are responsible for the contribution of approximately 80 per cent of the global gross domestic product. This has been attributed to diverse economic activities that are more feasible in urban than rural areas. For example, activities such as manufacturing, financial services, robust transportation services, tourism and entertainment that are associated with high economic returns are dominant in urban areas (M. Z. Allam, 2019; Z. Allam 2018, 2019b). In addition, the twenty-first-century phenomenon of rapid urban population growth and unprecedented urbanisation have contributed greatly in making cities the economic giants they are today. On the social front, cities have made an unmatched contribution in the growth of sectors such as health, education, entertainment and sport, hence providing urban residents an edge over their counterparts in rural areas.

On the flip side, cities have been in the spotlight for their contribution to the numerous global challenges being experienced today (Allam, 2020a, 2020f), for instance climate change, which is among the most pressing issues that the global community is experiencing, to which cities are major contributors. This can particularly be associated with the high affinity for energy and natural resources in cities – required to power sectors like the construction industry, transport, housing and manufacturing. The high consumption of these resources leads to an increased emission footprint, which when accumulated accounts for between 71 and 76 per cent of total global emissions (UN-Habitat, 2022). With climate change, different urban areas have experienced challenges like increased demand for infrastructural development that would guarantee resilience and adaptability. Further, there has been an unprecedented demand for infrastructure investment which can help mitigate negative consequences such as climate migration, where people move from affected areas to new habitats to be relatively safer and with increased opportunities (Barrios et al., 2006; International Organization for Migration, 2019; Thomas & Benjamin, 2018). Infrastructure development requirements in turn prompt an increased demand for financial resources, which have led to a sizeable number of cities plunging into debt (Batty, 2019).

Another global challenge that is directly associated with cities in the current century is that of socioeconomic inequality as expressed in Sustainable Development Goal (SDG) 11. The result of inequalities in cities includes increased urban poverty, homelessness, informal settlements and cases of insecurity. When all the aforementioned challenges are coupled with events such as pandemics (e.g. the recent outbreak of COVID-19), cities are left exposed and vulnerable, requiring urgent short-term and long-term interventions (Allam, 2020e). One viable pathway that can be explored in finding long-lasting solutions to some of those challenges is the increased adoption of existing and emerging smart technologies. In cities where they have already been deployed, there is already evidence that challenges like overreliance on non-renewable energy, traffic congestion and insecurity are slowly being addressed. Technologies have also been instrumental in critical decision making in areas like transport, health, housing and new economic frontiers.

Post-pandemic, it is anticipated that there will be an increased uptake of diverse technologies as people across the globe continue to adapt to new realities – like reduced human interaction and the urgency to revive and reinstate economies to pre-pandemic liveability levels (Allam & Dhunny, 2019; Allam et al., 2021, 2022b). The journey toward digitalisation, as was the case before COVID-19, will continue to be accelerated by the private sector and large corporations offering tech services and products (Amirtahasebi et al., n.d.; Global Infrastructure, 2021). Already tech companies, including Meta (formerly Facebook), have taken sides on what they anticipate the future of cities will look like (Facebook,

2021). Meta, together with other major corporations such as Microsoft and Apple, is individually exploring the possibility of introducing a new urban dimension dubbed the *metaverse* (Radoff, 2021a), where the physical and virtual realms will be seamlessly accessible courtesy of advanced technologies such as augmented reality, 6G, digital twins, 3D printing, robotics and cryptocurrency. The objectives in pursuing the futuristic urban planning models are to create more options for urban managers to pursue their quest to make cities more resilient against current and anticipated future challenges. Particularly, it is anticipated that the future deployment of different technologies will focus on creating urban environments that are more responsive to human-scale agendas, such as reduced travel, increased social activities and interactions, increased economic opportunities, reduced homelessness and reduced urban poverty.

Most of the conventional urban planning practices that have been dominant in the past have to some degree been aimed at integrating sustainability and inclusivity agendas in cities (Allam & Jones, 2021). In fact, some have even gained support from global policy frameworks such as the Paris Climate Accord and the SDGs, but unfortunately, they have not yielded sufficient outcomes in regard to solving social problems, nor in creating human-scale urban fabrics. This can be affirmed by recent reports showing that despite over 190 countries ratifying the Paris Agreement, the quest to prevent global temperatures rising above the targeted 2 degrees Celsius above pre-industrial levels is still unattainable (UNFCCC, 2021). This in part could be attributed to spirited human activities like overreliance on energy from non-renewable sources, increased demand for manufactured goods and increased migration to urban areas (Cleland, 1996). Going into the future, these and many other similar issues could be addressed by refocusing the urban planning agenda on social dimensions. This would, among other things, help to create urban environments where residents are provided with opportunities to take full control of their actions – like choosing whether to work remotely or in conventional workplaces. Further, the social dimension would allow residents to have diverse options to interact with family and friends (whether in the physical realm or in the virtual environment, especially in the case of the anticipated metaverse), providing alternative mobility options for neighbourhoods.

In light of the above background, this chapter will explore anticipated future urban planning models and trends emerging post-pandemic. It has also mapped the possible trends in the update of the different models, the benefits accrued and possible challenges ahead. The chapter further shows how existing and emerging technologies, coupled with externalities such as the impact of COVID-19 and anticipated increase in global temperatures as a result of climate change, will serve as harbingers for the uptake of new emerging planning models.

IS THE FUTURE OF CITIES IN AUTOMATION?

The emergence of COVID-19 and its subsequent impact on different urban spheres has prompted a renewed effort by different quarters to explore alternative options that could be adopted in restricting existing cities, as well as influence future cities. As has been widely shared in this book, the application of different technologies has been championed as a potential pathway that would help address some of the shortcomings that lead to many urban residents experiencing substantially more impacts than their counterparts in rural areas. In a bid to contain the wide spread of COVID-19, for instance, some of the most successful approaches that were instituted include advocating for social distancing and reduced travel – people were encouraged to work from home wherever possible (Chen, 2021; Glogowsky et al., 2021). In the urban planning spheres, such approaches might become mainstream, prompted by two factors. First, it has been observed that a sizeable number of urban residents are now apprehensive of being in spaces or engaging in activities that attract large crowds (Newsome, 2021). For instance, in almost all major sport events, there is evidence that fan attendance has not resumed to its previous levels and it is expected that it might take a long time before sport arenas regain their attraction (UKRI, 2021; PricewaterhouseCoopers, 2021). The major concern is catching the COVID-19 virus – the emergence of new variants affect even people who are fully vaccinated.

The second factor is related to efforts by stakeholders in urban planning realms such as information and communication technology (ICT) corporations and local governments to explore ways of ensuring that people resume their normal activities. However, that needs to be done in a way that allows people to avoid unwarranted social contact and excessive travel. Strategies that are being pursued to achieve this, and which will continue to increase in the future, include automation of different urban fabrics aimed at extending the capabilities and capacities of residents, urban infrastructures and local governments (Macrorie et al., 2021). The target of this is not only focused on reducing human interactions, but also ensuring that cities gain sufficient capacity to deal with other existing challenges like in the transport and environment sectors.

Automation in this context is being made possible by technologies such as artificial intelligence (AI) (Allam, 2020g), robotics, 3D printing, high-speed connectivity via 5G (Cugurullo, 2020; Murali, 2020; West, 2018) and the prospective 6G expected to be launched by 2030 (Mehta, 2020). Robotic technologies are slowly gaining traction in sectors like the construction industry, thus allowing increased output, efficiency, reduced labour costs and the emergence of new construction innovations and platforms. For instance, using robotics, coupled with other technologies, it is now possible to pre-fabricate housing units in factories, hence reducing the time taken for residential projects to be completed. This further helps in reducing the overall cost of production and promotes environmentally sustainable practices by reducing excessive consumption of natural resources (Pan et al., 2018). Whereas there is some genuine criticism of adopting this technology, especially with regard to accelerating unemployment rates (Cameron, 2020; Davidson, 2017; West, 2018), it is anticipated that in the future, jobs taken by automation will be compensated by the emergence of new economic frontiers (Macrorie et al., 2019; Manyika et al., 2017). These will increase exponentially after the launch of the metaverse concept, which proponents argue will create numerous opportunities for designers, programmers and other groups drawn from the technology realm (Hackl, 2021; Radoff, 2021b).

During the height of COVID-19, especially in the period when people were 'trapped' in lockdowns, some countries were reported to have deployed the use of digital products such as drones and robots to not only enforce health protocols, but also to deliver vital supplies like personal protective equipment and medicine to known COVID-19 hotspots (Allam, 2020c, 2020d; Chen et al., 2020; Gulistan News, 2020). These strategies are hailed to have been effective in preventing interpersonal transmission as well as helping overcome the constraints prompted by reduced transportation activities (Yang & Reuter, 2020). In cities, these technologies are expected to continue being deployed, including in other sectors like logistics and security, to help increase efficiency and productivity (Galer, 2020). Proponents of the technology believe that the increasing amount of data being generated by different urban fabrics will make it possible to deploy the technology in remote areas like hazardous recycling plants, making it possible to not only dispose of harmful products in a safe manner, but also to help in recycling parts of those products (Simpson, 2016), thus promoting the concept of the circular economy.

Proponents of urban planning models such as the smart city concept anticipate that the exponential advancement and emergence of new smart technologies and innovative concepts such as the smart city brain, metaverse and others are likely to prompt cities to become autonomous. This will play a significant role in the morphology of cities and will prompt most activities to be automated, hence requiring little or no human intervention (Cugurullo, 2020), ultimately reverberating on social dimensions and provision of infrastructures. As noted above, this will be critical in sectors like transport, energy, waste management and others that have traditionally contributed greatly to emissions. The premise made by many researchers is that most of these sectors will experience exponential advancement with products such as autonomous cars, smart grids, smart meters and smart houses, to name a few, becoming an intrinsic part of the city. The 'intelligence' levels of different urban products such as cars and robots are expected to supersede existing technologies, especially as most of these products will be powered by technologies such as artificial neural networks that are fashioned to mimic the human brain (Allam, 2019a).

Despite the many positive innovations and creative outcomes that are anticipated to emerge in cities in the short- and long-run, there are fears that they might attract substantial negative impacts that demand to be addressed before most innovations become functional. For instance, many critics are concerned about the ethics of AI in cities, especially due to its influence that will prompt cities to morph into autonomous entities able to self-regulate, probably in a manner different from how humans would perform. Other ethical concerns pertain to technologies replacing human beings, especially in job markets, thus rendering substantial urban populations jobless (Bostrom, 2017), especially in cities revolved around manufacturing cores. While there are promises that future cities might present new job opportunities both in the physical and virtual realms, replacing human beings in existing positions does not sit well, especially considering current and anticipated economic hardships. Further, it is projected that the transition from current urban setups to new emerging models like 15-minute cities, the metaverse and others will require substantial financial investment. However, following the impacts of climate change and the pandemic, cities have already experienced unprecedented financial challenges, and it might prompt moral and ethical concerns if more resources are directed toward implementing new planning concepts when the global population is still trying to recover from the aftermath of the aforementioned challenges.

THE COVID-19 PANDEMIC AS A CATALYST TO AUTONOMOUS CITIES

The emergence of the fourth industrial revolution in the mid-2000s prompted a significant boost to cities across the globe, more so in catalysing the adoption of data-driven approaches in urban planning. The quest to have cities' decision-making processes based on data began as early as the 1970s in Los Angeles but gained traction very slowly. It was only after Industry 4.0 emerged that an exponential increase of the number of cities embracing technologically oriented approaches to urban planning and administration was observed (Brasuell, 2015). By 2011, approximately 24 cities across the world had taken up the challenge and implemented some form of smart city initiative spearheaded by large ICT firms like IBM and Cisco (Söderström et al., 2014). The popularity of this concept grew rapidly such that by the end of 2019, the number of cities across the globe (mostly in developed and some developing economies) that had adopted the use of diverse smart technologies had increased to approximately 102 (IMD, 2019). Further, it is reported that many more cities have contemplated embracing this growing trend, and were it not for factors like financial constraints, especially in developing and poorer countries, the number of smart cities would have been significantly higher. It is expected that the new realities prompted by COVID-19 will significantly push the number of those already embracing this planning model to increase rapidly.

The COVID-19 pandemic emerged in early 2020 and subsequently prompted lockdowns during the first and second quarters of that year, hence, the need for connectivity warranted by the adoption of technology became apparent (Allam, 2020b, 2020c, 2020d). For most cities that had initiated some form of 'smartness', residents are reported to have had some relief in terms of work, schooling, communication and accessing vital information (Boza-Kiss et al., 2021; Sharifi & Khavarian-Garmsir, 2020). This was totally different in areas that had little or no technology deployment. For instance, it took extended periods of time for learners to resume learning activities in a sizeable number of countries (mostly in developing and less developed economies) (UNICEF, 2021). Similarly, while it was possible to implement working from home, conduct virtual meetings and conduct online financial transactions and e-commerce in urban areas located in developed and some developing economies, people in the least developed economies had no such 'luxuries' (Mathrani et al., 2021). This highlighted the disparities that exist between cities in developed and less developed economies.

The above background is important as it paints a picture of the current global situation in urban areas, and also helps to point out critical areas to which attention should be focused to ensure that the global community is sufficiently equipped to overcome similar pandemics in the future. From a historical perspective, cities always bounce back from pandemics and crises, and in most cases on the crest of rapid economic growth. However, the recovery is not always smooth, especially for those without solid infrastructural investments. Therefore, regarding the COVID-19 pandemic and its anticipated global economic recovery plans, there will be an urgent need to prioritise investing in technologies that could, among other things, help automate urban activities, leading to increased efficiency and performance (Allam & Jones, 2021). Technologies will further help urban managers create opportunities for alternative options to be explored in different sectors like transport, energy, health, education and environmental sustainability. Focusing on technology post-pandemic will also be critical in addressing the challenges of climate change, which experts predict might be accelerated as countries increase their focus on economic recovery (Allam, 2020e). During the height of COVID-19, reports show that there were significant declines in emissions in cities due to reduced economic and social activities. This could be accelerated using technologies, especially when attention is redirected toward exploiting alternatives in sectors like transport where it is possible to adopt diverse mobility options.

Besides playing a proactive role in helping achieve climate targets set for 2030, cities also have their objectives on socioeconomic inclusivity well defined. In SDG 11, it is clearly spelt out that cities and human settlements need to be inclusive, safe, resilient and sustainable. While urban planning models such as smart cities, compact cities and sustainable cities that gained popularity due to the adoption of technologies in cities have been instrumental in achieving positive outcomes toward those objectives, there is much that still needs to be achieved. The popularisation of the '15-minute city' concept has the capacity to help address some of those objectives, especially in relation to social dimensions (Allam et al., 2022a, 2022b; Moreno et al., 2021; Weng et al., 2019). However, those social dimensions need to be complemented by also concentrating on the environmental and economic dimensions. On the economic frontier, a concerted effort would need to be directed toward creating more employment opportunities to compensate for the jobs lost during the pandemic, as well as those that have been rendered obsolete by digitalisation and the automation of different urban spheres. With respect to sustainability, extra attention will need to be directed toward revolutionising the energy sector, making sure that sufficient resources are available for renewable energy options. Further, it will be necessary to ensure that stakeholders capitalise on technology use to invest in smart and green products in the housing (e.g. smart houses, green spaces) and transport sectors (e.g. electric vehicles, use of bicycles, smart parking and others) to continue to accelerate the achievement of sustainability agendas. This will not only allow for the achievement of last-mile connectivity, especially in the least developed economies where most informal sectors abound (Bosworth et al., 2020; Kåresdotter et al., 2022), but will also help in reducing the cost of investing in those technologies.

Renewing the agenda on sustainability, especially by emphasising decarbonisation in cities, as envisioned in COP 26, will be critical, especially on the micro- and macro-levels (United Nations Framework Convention on Climate Change, 2021). On the macro level, this will help safeguard the plight of vulnerable economies such as small island developing states, which have experienced unprecedented challenges over the years following economic activities taking place in areas outside their jurisdiction (UNEP, 2014). These economies, together with others in low-lying and coastal areas, have experienced numerous challenges, including incurring huge debts as they pursue mitigation programmes that guarantee resilience and adaptability (Batty, 2019; Kling et al., 2018). Therefore, actions taken in cities across the globe post-pandemic will not only impact local economies, but will also have significant impacts on the global scale.

POSSIBLE EMERGING TECHNOLOGIES AND THE NEED FOR URBAN POLICY CHANGE

Technology development in the past decades has had a profound influence and impact on the different facets of humanity and it is expected that this will escalate going into the future. In the urban planning arena, technology development has helped to resolve numerous challenges that would have been unsurmountable if traditional approaches were to be relied upon. For instance, with regard to accessibility of different nodes within urban areas, technology has played a critical role, with respect to roads, railways, water and air transportation development and improvements. Indeed, it is safe to argue that the unprecedented growth in the urban population could be partly credited to growth in the transport sector, by making it possible for people and products to be efficiently moved from one location to another in record time, safely and at a relatively affordable cost. Essentially, it has been argued that most conventional urban planning models were fashioned to support the flow of traffic, thus affirming the importance of this sector in the growth of cities. Another sector that has greatly benefitted from the increased development of technology is the communication sector. Due to advancements in technology, efficient platforms such as the internet, smart mobile devices, social media and other technologically oriented methodologies have revolutionised how people communicate, share and receive information. Courtesy of development in this technology, urban areas are now being designed, planned and governed using data generated from diverse communication products (Bibri & Krogstie, 2020; Nikitin et al., 2016; Sutherland & Cook, 2017).

Going into the future, it is anticipated that more technologies with the potential to revolutionise different aspects of humanity are expected to emerge, in order to advance the successes that have already been achieved. For instance, to operationalise wireless communication further, thus allowing for high-speed connectivity and the real-time collection and transfer of data, it is anticipated that 6G will emerge to replace 5G that is currently gaining traction in many cities (Samsung, 2020). With the 5G technology, a number of cutting-edge technologies and innovations have already emerged, and proponents of 6G are of the view that numerous new offshoot technologies will be developed. Some of those, already being conceptualised, include comprehensive AI, immersive reality and 4D printing. These will be inspired by the hyperconnectivity speed (over 1 terabyte per second) that is expected to be achieved with 6G, plus the low latency that proponents are looking forward to in this next generation of wireless connectivity (Tariq et al., 2020). Theoretically, it could be argued that the 6G technology will prompt challenges like high energy demand, thus defeating the main objective of upgrading existing wireless connectivity. However, it is believed that existing technologies such as AI and machine learning will help in optimising the infrastructure and networks supporting 6G to guarantee even lower energy consumption (IEEE Communications Society, 2021).

Another technology that is expected to gain extra ground in the coming years is digital twins, which in broad terms can be defined as the representation of replicas of physical objects or systems in virtual environments. Virtual objects are then updated continually and in real time, hence allowing remote observation, simulation, test, diagnosis and decision making on physical objects (Armstrong, 2020). This technology has already been tested in the manufacturing industries, especially by auto companies and airplane manufacturers, and this has allowed them to cut costs in areas like testing where they no longer need to make models to be deployed for various tests, fleet management, design improvement and troubleshooting among other things (Altexsoft, 2021; Rudskoy et al., 2021; Zheng et al., 2019). In urban planning, this is expected to prompt a big impact in helping to design and make cities that are responsive to environmental, social and economic demands. It is expected that this technology will also benefit from a wide range of smart devices and sensors that are already installed, and those that will continue to be deployed in urban infrastructure (Horwitz, 2019).

Another technology that is expected to gain momentum in the future, especially in support of remote working, reduced demand for automobiles and increased uptake of cultural heritage and creative art conservation, is augmented reality. The whole range of virtual reality (VR) technologies such as augmented reality, mixed reality and VR 360^0 has been gaining traction in different fields like manufacturing, the health sector, the entertainment sector, sports and travel. But it is anticipated that these will become even more immersive in the future, especially with the emergence of other advanced technologies. These are expected to revolutionise sectors in which they are already being deployed as well as help in the actualisation of innovations such as the metaverse that is being conceptualised by big ICT technology giants such as Meta (Facebook, 2021), Microsoft and Apple (Radoff, 2021a). The metaverse is expected to utilise both existing and emerging technologies to allow global citizens to experience a new dimension based on immersive reality, digital twins, cryptocurrency, AI and other technologies. These technologies will further allow products such as autonomous vehicles, unmanned aerial vehicles, robots and 3D printers to become mainstream in different urban spheres. In sectors like the construction industry, the pressure exerted by an increasing urban population will prompt innovations such as the mass production of module houses that will be relatively cheap and responsive to sustainability agendas.

Emerging technologies and innovations are expected to bring about new challenges to urban managers, hence the need for urban policy changes to guide and control their application. Already with existing technologies there have been a number of issues, including financial demands, socioeconomic inequalities and security and privacy concerns, that have triggered ethical and moral issues of technology use in urban areas (Kitchin, 2016; Mark & Anya, 2019). A majority of those concerns have been addressed using different urban policies, but some like security and privacy, especially of massive data being collected, still linger. The rise of new dimensions such as the metaverse will prompt a need for new policies to guide different issues like privacy, security and ownership of data, the nature of virtual trading, health concerns of users and equity (Dick, 2021).

COULD THE ANTICIPATED FUTURE SMART CITIES BRING ABOUT MORE URBAN CHALLENGES?

The increasing growth and development of smart cities across the globe is expected to attract immeasurable benefits for different urban dimensions. However, an almost equal measure of challenges is expected to arise. Some of those challenges may become complex as different externalities like the war between Russia and Ukraine (Bond et al., 2022) and the prevailing impacts of COVID-19 (Kose et al., 2020; Walton & Aalst, 2020), result in full-blown economic crisis (World Bank, 2020). The challenges will further be complicated by the ever growing urban population (expected to reach a high of over 60 per cent by 2030 and 70 per cent by 2050) (Kaneda et al., 2020). Already, cities that have taken bold strides toward becoming 'smart' are experiencing unique challenges like financial constraints, lack of sufficient legal and technical frameworks to guide the implementation of the projects, security and privacy concerns and numerous impacts of climate change, among others.

Before the emergence of the COVID-19 pandemic, the aforementioned challenges were anticipated, and steps toward countering and mitigating a majority of those were being taken. For instance, with regard to climate change, global policies were formulated as showcased in documents such as the Paris Climate Agreement (UNFCCC, 2015), the 2030 Agenda for Sustainable Development (Government of Mauritius, 2017), the New Urban Agenda and the SDGs (United Nations Development Programme, 2015). Each of these proposes comprehensive policy frameworks that could help tilt the weight of the climate change discourse in cities. With regard to financial constraints, many cities across the globe had crafted strategies and methodologies to facilitate smooth and gradual financial support to different smart projects (Hamilton & Zhu, 2017). These included collaborations between public and private sectors in public-private partnership arrangements (APMG International, 2016), debt financing, etc. On privacy and security, there is evidence that efforts such as robust public participation, policy formulation and capacity building in local governments were being emphasised to allow for close monitoring and management of data and physical smart city infrastructures (Raco, 2009; van Zoonen, 2016).

Post-pandemic, however, it is anticipated that some of these challenges, despite having been partly addressed, might escalate, while new ones might also emerge. On the escalation, despite the forward steps made in reducing emissions, more so in urban areas, the urge to expedite economic recovery in different countries has prompted a resurgence of unsustainable practices like the use of fossil fuels in manufacturing sectors (Plumer & Popovich, 2018; United Nations Environment Programme, 2021). As of 2021, the rebound on carbon emissions had reached 4.9 per cent after experiencing a 5.6 per cent reduction in 2020, courtesy of COVID-19-related mitigation protocols (United Nations Environment Programme, 2021). The rise in emissions then would translate to more climate change incidences, which in turn would prompt long-lasting impacts on cities. Such incidences have the potential to prompt the destruction of critical smart city infrastructures, thus increasing the initial capital requirement for these projects to be completed.

One of the blueprint objectives of the smart city concept is to improve the liveability status in urban areas by promoting 'smart' practices like the use of alternative mobility options and the adoption of cleaner energy including housing and automobiles (Yigitcanlar et al., 2018). However, the liveability agenda might not be inclusive, especially following the resurgence of the use of fossil fuels plus the long-term impacts of the Russia-Ukraine war putting a strain on European cities as well as others around the world (Bond et al., 2022; Le Quéré et al., 2021; Leahy, 2019). In particular, projections are that the conflict will have ripple effects on the cost of living and prompt an increase in inflation, especially of food (Nicas, 2022). Further, unless alternative energy sources are explored and adopted in Europe and other regions, like North America, which is also experiencing energy shortages following the embargo placed on Russia, most residents of future cities will have to live without heating services (Bond et al., 2022). In cases where heating services are available, this will be at a premium, which most urban residents, especially those in the lower-middle and lower-income cadres, might find too expensive to afford. Therefore, the liveability promise will remain a pipedream until the situation is rescued by investing in renewable energies.

Implementing smart city projects – from planning, design, implementation and maintenance – is an expensive undertaking, and most cities do not have sufficient capacity to finance this from their public resources. Turning to the private sector for partnerships, or securing external loans, has also proven expensive, with cities risking critical public assets. Further, it is urban residents who ultimately shoulder the burden of repaying these loans through taxation (Nishio, 2021). In the future, with more complex and advanced technologies anticipated to emerge, the cost of securing and deploying those technologies will be relatively higher than it is today, hence it is probable that different taxation mechanisms might become more expensive. That said, the reprieve for urban residents will be pegged on the quality and quantity of opportunities that will be generated after the projects are implemented.

On the issue of privacy and security of both personal data and the physical infrastructures that support most urban systems, expectations are that policies and legal frameworks will be in place. However, looking at the current scenario, there are loopholes being exploited, thus making people apprehensive about what the future might hold (Allam, 2019c; Ismagilova et al., 2020; van Zoonen, 2016). From the literature, the subject of privacy and security could be addressed amicably by expediting collaboration between all stakeholders in the urban arena. However, the potential that data have in terms of earning competitiveness might pose a hindrance to the formulation of comprehensive data management platforms and laws (Franke & Gailhofer, 2021). Another challenge related to data and privacy that might befall cities is the complexity of systems, noting that new technologies such as 6G will have become mainstream, and new dimensions like the metaverse, which might not be within the jurisdiction of local governments, will also be in play. It will be difficult to control and monitor the virtual trading of property and other transactions, as ownership of metaverse spaces will be beyond local governments' control (Roy, 2021), and laws and regulations controlling them might not be available by the time people start exploring this dimension.

While the smart city planning model poses a potent solution for increasing the efficiency and performance of cities, there are numerous uncertainties that need to be addressed to render safer, more sustainable, inclusive and resilient cities.

REFERENCES

Alex, B. (2020, 28 August). Which ancient city is considered the oldest in the world? Discover. Retrieved 23 March 2022 from www.discovermagazine .com/planet-earth/which-ancient-city-is-considered-the-oldest-in-the-world

Allam, M. Z. (2019). *Urban Resilience and Economic Equity in an Era of Global Climate Crisis*. University of Sydney.

Allam, Z. (2018). Contextualising the smart city for sustainability and inclusivity. *New Design Ideas*, *2*(2), 124–127.

Allam, Z. (2019a). Achieving neuroplasticity in artificial neural networks through smart cities. *Smart Cities*, *2*(2). https://doi.org/10.3390/smartcities2020009

Allam, Z. (2019b). *Cities and the Digital Revolution: Aligning Technology and Humanity*. Springer Nature.

Allam, Z. (2019c). The emergence of anti-privacy and control at the nexus between the concepts of safe city and smart city. *Smart Cities*, *2*(1), 96–105.

Allam, Z. (2020a). Religious matrimony, urban sprawl and urban morphology. In *Theology and Urban Sustainability* (pp. 21–35). Springer.

Allam, Z. (2020b). The first 50 days of COVID-19: A detailed chronological timeline and extensive review of literature documenting the pandemic. *Surveying the COVID-19 Pandemic and Its Implications*, *1*, 1–7, doi: 10.1016/ B978-0-12-824313-8.00001-2.

Allam, Z. (2020c). The second 50 days: A detailed chronological timeline and extensive review of literature documenting the COVID-19 pandemic from day 50 to day 100. *Surveying the COVID-19 Pandemic and Its Implications*, *9*.

Allam, Z. (2020d). The third 50 days: A detailed chronological timeline and extensive review of literature documenting the COVID-19 pandemic from day 100 to day 150. *Surveying the COVID-19 Pandemic and Its Implications*, *41*.

Allam, Z. (2020e). Vital COVID-19 economic stimulus packages pose a challenge for long-term environmental sustainability. In *Surveying the COVID-19 Pandemic and Its Implications* (pp. 97–105). Elsevier.

Allam, Z. (2020f). Urban and graveyard sprawl: The unsustainability of death. In *Theology and Urban Sustainability* (pp. 37–52). Springer.

Allam, Z. (2020g). Urban chaos and the AI Messiah. In *Cities and the Digital Revolution* (pp. 31–60). Palgrave Pivot.

Allam, Z., & Dhunny, A. Z. (2019). On big data, artificial intelligence and smart cities. *Cities*, *89*, 80–91.

Allam, Z., & Jones, D. S. (2021). Future (post-COVID) digital, smart and sustainable cities in the wake of 6G: Digital twins, immersive realities and new urban economies. *Land Use Policy*, *101*, 105201.

Allam, Z., Jones, D., Biyik, C., Allam, Z., & Takun, Y. R. J. R. i. G. (2021). Rewriting city narratives and spirit: Post-pandemic urban recovery mechanisms in the shadow of the global 'black lives matter' movement. *Research in Globalization*, *3*, 100064.

Allam, Z., Nieuwenhuijsen, M., Chabaud, D., & Moreno, C. (2022a). The 15-minute city offers a new framework for sustainability, liveability, and health. *The Lancet Planetary Health*, *6*(3), e181–e183.

Allam, Z., Bibri, S. E., Jones, D. S., Chabaud, D., & Moreno, C. (2022b). Unpacking the '15-minute city' via 6G, IoT, and digital twins: Towards a new narrative for increasing urban efficiency, resilience, and sustainability. *Sensors, 22*(4). https://doi.org/10.3390/s22041369

Altexsoft. (2021, 16 September). Digital twins: Components, use cases, and implementation tips. Altexsoft. Retrieved 4 January 2022 from www.altexsoft .com/blog/digital-twins/

Amirtahasebi, R., Orloff, M., & Wahba, S. (n.d.). Partnering arrangements with the private sector. World Bank. Retrieved 22 April 2021 from https://urban -regeneration.worldbank.org/node/87

APMG International. (2016). How a private finance PPP project is financed: Where the money to pay construction costs comes from. In *PPP Certification Guide*. World Bank Group.

Armstrong, M. M. (2020, 4 December). Cheat sheet: What is digital twin? IBM. Retrieved 4 January 2022 from www.ibm.com/blogs/internet-of-things/iot -cheat-sheet-digital-twin/

Barrios, S., Bertinelli, L., & Strobl, E. (2006). Climate change and rural-urban migration: The case of sub-Saharan Africa. *Journal of Urban Economics, 60*(3), 357–371.

Batty, M. (2019). Cities in debt. *Environment and Planning B: Urban Analytics and City Science, 46*(2), 203–206.

Bibri, S., & Krogstie, J. (2020). Data-driven smart sustainable cities of the future: A novel model of urbanism and its core dimensions, strategies, and solutions. *Journal of Futures Studies*. https://doi.org/10.6531/JFS.202012_25(2).0009

Bond, I., Cornago, E., Mortera-Martinez, C., & Scazzieri, L. (2022, 17 March). Russia's war on Ukraine: There is worse to come (for the West as well). Centre for European Reform. Retrieved 26 March 2022 from www.cer.eu/insights/ russias-war-ukraine-worse-west

Bostrom, N. (2017). *Superintelligence*. Oxford University Press.

Bosworth, G., Price, L., Collison, M., & Fox, C. (2020). Unequal futures of rural mobility: Challenges for a 'smart countryside'. *Local Economy, 35*(6), 586–608.

Boza-Kiss, B., Pachauri, S., & Zimm, C. (2021). Deprivations and inequities in cities viewed through a pandemic lens. *Frontiers, 3*. https://doi.org/10.3389/ frsc.2021.645914

Brasuell, J. (2015, 22 June). The early history of the 'smart cities' movement – in 1974 Los Angeles. Planetizen. Retrieved 20 January 2021 from www .planetizen.com/node/78847

Cameron, E. (2020). How will automation impact jobs? PWC. Retrieved 10 March 2020 from www.pwc.co.uk/services/economics/insights/the-impact-of -automation-on-jobs.html

Chen, B., Marvin, S., & While, A. (2020). Containing COVID-19 in China: AI and the robotic restructuring of future cities. *Dialogues in Human Geography, 10*(2), 238–241.

Chen, Z. (2021). Influence of working from home during the COVID-19 crisis and HR practitioner response. *Frontiers in Psychology, 12*. www.frontiersin .org/article/10.3389/fpsyg.2021.710517

Cleland, J. (1996). Population growth in the 21st century: Cause for crisis or celebration? *Tropical Medicine and International Health, 1*(1), 15–26.

Cugurullo, F. (2020). Urban artificial intelligence: From automation to autonomy in the smart city. *Frontiers in Sustainable Cities, 2*. www.frontiersin.org/article/10.3389/frsc.2020.00038

Davidson, P. (2017, 1 December). Automation could kill 73 million US jobs by 2030. Transport Topics. Retrieved 10 March 2020 from www.ttnews.com/articles/automation-could-kill-73-million-us-jobs-2030

Dick, E. (2021, 15 November). Public policy for the metaverse: Key takeaways from the 2021 AR/VR Policy Conference. Information Technology and Innovation Foundation. Retrieved 24 March 2021 from https://itif.org/publications/2021/11/15/public-policy-metaverse-key-takeaways-2021-arvr-policy-conference

Facebook. (2021, 28 October). Connect 2021: Our vision for the metaverse. Meta. Retrieved 1 December 2021 from https://tech.fb.com/connect-2021-our-vision-for-the-metaverse/

Franke, J., & Gailhofer, P. (2021). Data governance and regulation for sustainable smart cities. *Frontiers in Sustainable Cities, 3*. www.frontiersin.org/article/10.3389/frsc.2021.763788

Galer, S. (2020, 17 November). Beyond smart cities: How drones and robotics revolutionize building inspections. Forbes. Retrieved 24 March 2022 from www.forbes.com/sites/sap/2020/11/17/beyond-smart-cities-how-drones-and-robotics-revolutionize-building-inspections/?sh=885f65b10c2a

Global Infrastructure, H. (2021). Private sector roles and participation. Global Infrastructure Hub. Retrieved 15 April 2021 from https://inclusiveinfra.gihub.org/action-areas/private-sector-roles-and-participation/

Glogowsky, U., Hansen, E., & Schächtele, S. (2021). How effective are social distancing policies? Evidence on the fight against COVID-19. *PLOS ONE, 16*(9), e0257363.

Government of Mauritius. (2017, 31 October). Agenda 2030 an effective guideline for Mauritius to achieve new heights, reiterates PM. Government of Mauritius. Retrieved 5 March 2021 from www.govmu.org/English/News/Pages/Agenda-2030-an-effective-guideline-for-Mauritius-to-achieve-new-heights,-reiterates-PM-.aspx

Gulistan News. (2020, 20 April). Chile: A drone delivers medicine to the elderly who are socially isolated to prevent contracting the coronavirus, Zapallar, Chile, Sunday, Twitter. Retrieved 2 February 2022 from https://twitter.com/GulistanNewsIn/status/1252232953788198915

Hackl, C. (2021, 15 March). Making money in the metaverse. Forbes. Retrieved 1 December 2021 from www.forbes.com/sites/cathyhackl/2021/03/15/making-money-in-the-metaverse/?sh=13e7c60e3b43

Hamilton, S., & Zhu, X. (2017). Funding and financing smart cities. Retrieved 2 February 2022 from www2.deloitte.com/content/dam/Deloitte/us/Documents/public-sector/us-ps-funding-and-financing-smart-cities.pdf

Horwitz, L. (2019, 19 July). The future of IoT miniguide: The burgeoning IoT market continues. Cisco. Retrieved 15 January 2020 from www.cisco.com/c/en/us/solutions/internet-of-things/future-of-iot.html

IEEE Communications Society. (2021, November). AI and 6G convergence: An energy-efficiency perspective. IEEE Com Soc. Retrieved 24 March 2022 from www.comsoc.org/publications/magazines/ieee-network/cfp/ai-and-6g -convergence-energy-efficiency-perspective

IMD. (2019, October). IMD smart city index 2019. IMD. Retrieved 23 March 2022 from www.imd.org/research-knowledge/reports/imd-smart-city-index -2019/#:~:text=The%20IMD%20World%20Competitiveness%20Center,which %20ranks%20102%20cities%20worldwide

International Organization for Migration. (2019). *Climate Change and Migration in Vulnerable Countries: A Snapshot of Least Developed Countries, Landlocked Developing Countries and Small Island Developing States*. International Organization for Migration.

Ismagilova, E., Hughes, L., Rana, N. P., & Dwivedi, Y. K. (2020). Security, privacy and risks within smart cities: Literature review and development of a smart city interaction framework. *Information Systems Frontiers*. https://doi .org/10.1007/s10796-020-10044-1

Kaneda, T., Greenbaum, C., & Kline, K. (2020). 2020 world population data sheet. Report. PRB. Retrieved 1 February 2022 from www.prb.org/wp -content/uploads/2020/07/letter-booklet-2020-world-population.pdf

Kåresdotter, E., Page, J., Mörtberg, U., Näsström, H., & Kalantari, Z. (2022). First mile/last mile problems in smart and sustainable cities: A case study in Stockholm County. *Journal of Urban Technology*, *29*(2), 115–137.

Kitchin, R. (2016). The ethics of smart cities and urban science. *Philosophical Transactions of the Royal Society A: Mathematical, Physical and Engineering Sciences*, *374*(2083), 1–15.

Kling, G., Lo, Y., Murinde, V., & Volz, U. (2018). Climate vulnerability and the cost of debt. *SSRN Electronic Journal*, 1–30. https://doi.org/10.2139/ssrn .3198093

Kose, M. A., Ohnsorge, F., Nagle, P., & Sugawara, N. (2020). Caught by the cresting debt wave. *Finance and Development*, *57*(2), 40–43.

Le Quéré, C., Peters, G. P., Friedlingstein, P., Andrew, R. M., Canadell, J. G., Davis, S. J., Jackson, R. B., & Jones, M. W. (2021). Fossil CO2 emissions in the post-COVID-19 era. *Nature Climate Change*, *11*(3), 197–199.

Leahy, S. (2019, 5 November). Most countries aren't hitting 2030 climate goals, and everyone will pay the price. *National Geographic*. Retrieved 7 August 2021 from www.nationalgeographic.com/science/article/nations-miss-paris -targets-climate-driven-weather-events-cost-billions

Macrorie, R., Marvin, S., & While, A. (2019). Robotics and automation in the city: A research agenda. *Urban Geography*, 1–21. https://doi.org/10.1080/ 02723638.2019.1698868

Macrorie, R., Marvin, S., & While, A. (2021). Robotics and automation in the city: A research agenda. *Urban Geography*, *42*(2), 197–217.

Manyika, J., Lund, S., Chui, M., Bughin, J., Woetzel, J., Batra, P., Ko, R., & Sanghv, S. (2017). Jobs lost, jobs gained: workforce transitions in a time of automation. McKinsey. Retrieved 2 February 2022 from www.mckinsey.com/~/ media/mckinsey/industries/public%20and%20social%20sector/our%20insights/ what%20the%20future%20of%20work%20will%20mean%20for%20jobs

%20skills%20and%20wages/mgi-jobs-lost-jobs-gained-executive-summary
-december-6-2017.pdf

Mark, R., & Anya, G. (2019). Ethics of using smart city AI and big data: The case of four large European cities. *The ORBIT Journal, 2*(2), 1–36.

Mathrani, A., Sarvesh, T., & Umer, R. (2021). Digital divide framework: Online learning in developing countries during the COVID-19 lockdown. *Globalisation, Societies and Education*, 1–16. https://doi.org/10.1080/14767724.2021.1981253

Mehta, S. S. (2020, 20 June). Mission 6G: Time to lead. *The Tribune*. Retrieved 21 July 2020 from www.tribuneindia.com/news/comment/mission-6g-%E2%80%94-time-to-lead-115277

Moreno, C., Allam, Z., Chabaud, D., Gall, C., & Pratlong, F. (2021). Introducing the '15-minute city': Sustainability, resilience and place identity in future post-pandemic cities. *Smart Cities, 4*(1), 93–111.

Murali, S. (2020, January). Automation makes it possible for cities to go green. Smart Cities Dive. Retrieved 10 July 2020 from www.smartcitiesdive.com/news/automation-makes-it-possible-for-cities-to-go-green/569672/

Newsome, M. (2021, 3 May). 'Cave syndrome' keeps the vaccinated in social isolation. *Scientific American*. Retrieved 24 March 2022 from www.scientificamerican.com/article/cave-syndrome-keeps-the-vaccinated-in-social-isolation1/

Nicas, J. (2022, 20 March). Ukraine war threatens to cause a global food crisis. *New York Times*. Retrieved 25 March 2022 from www.nytimes.com/2022/03/20/world/americas/ukraine-war-global-food-crisis.html

Nikitin, K., Lantsev, N., Nugaev, A., & Yakavleva, A. (2016). Data-driven cities: From concept to applied solutions. PricewaterhouseCoopers. Retrieved 2 February 2022 from http://docplayer.net/50140321-From-concept-to-applied-solutions-data-driven-cities.html

Nishio, A. (2021, 10 August). Facing substantial investment needs, developing countries must sustainably manage debt. Retrieved from https://blogs.worldbank.org/voices/facing-substantial-investment-needs-developing-countries-must-sustainably-manage-debt

Pan, M., Linner, T., Cheng, H. M., Pan, W., & Bock, T. (2018). A framework for utilizing automated and robotic construction for sustainable building. Proceedings of the 21st International Symposium on Advancement of Construction Management and Real Estate, Singapore.

Plumer, B., & Popovich, N. (2018, 7 December). The world still isn't meeting its climate goals. *New York Times*. Retrieved 12 June 2019 from www.nytimes.com/interactive/2018/12/07/climate/world-emissions-paris-goals-not-on-track.html

PricewaterhouseCoopers. (2021). The future of sports infrastructure in a post-pandemic world. PWC. Retrieved 23 March 2022 from www.pwc.co.uk/who-we-are/regional-sites/london/insights/future-of-sports-infrastructure-in-post-pandemic-world.html

Raco, M. (2009). Governance, urban. In R. Kitchin & N. Thrift (Eds), *International Encyclopedia of Human Geography* (pp. 622–627). Elsevier.

Radoff, J. (2021a, 12 November). Clash of the metaverse titans: Microsoft, Meta and Apple. Medium. Retrieved 1 December 2021 from https://medium.com/building-the-metaverse/clash-of-the-metaverse-titans-microsoft-meta-and-apple-ce505b010376

Radoff, J. (2021b, 11 October). Jobs in the metaverse. Medium. Retrieved 7 December 2021 from https://medium.com/building-the-metaverse/jobs-in-the-metaverse-9395db90086

Roy, A. (2021, 2 December). Doing business in the metaverse: Opportunity or Threat? XR Today. Retrieved 4 December 2021 from www.xrtoday.com/virtual-reality/doing-business-in-the-metaverse-opportunity-or-threat/

Rudskoy, A., Ilin, I., & Prokhorov, A. (2021). Digital twins in the intelligent transport systems. *Transportation Research Procedia, 54*, 927–935.

Samsung. (2020). *6G:* The next hyper-connected experience for all. Samsung Research. Retrieved 21 July 2020 from https://cdn.codeground.org/nsr/downloads/researchareas/6G%20Vision.pdf

Sharifi, A., & Khavarian-Garmsir, A. R. (2020). The COVID-19 pandemic: Impacts on cities and major lessons for urban planning, design, and management. *Science of the Total Environment, 749*, 142391.

Simpson, W. (2016, 2 September). The rise of the recycling robots. Resource. Retrieved 24 March 2022 from https://resource.co/article/rise-recycling-robots-11340

Söderström, O., Paasche, T., & Klauser, F. (2014). Smart cities as corporate storytelling. *City, 18*(3), 307–320.

Sutherland, M. K., & Cook, M. E. (2017). Data-driven smart cities: A closer look at organizational, technical and data complexities. Proceedings of the 18th Annual International Conference on Digital Government Research, Staten Island, New York.

Tariq, F., Khandaker, M. R., Wong, K.-K., Imran, M. A., Bennis, M., & Debbah, M. J. I. W. C. (2020). A speculative study on 6G. *IEEE, 27*(4), 118–125.

Thomas, A., & Benjamin, L. (2018). Policies and mechanisms to address climate-induced migration and displacement in Pacific and Caribbean small island developing states. *International Journal of Climate Change Strategies and Management, 10*(1), 86–104.

UKRI (2021, 21 August). What's the future of live events post-pandemic. UKRI. Retrieved 23 March 2022 from www.ukri.org/news-and-events/tackling-the-impact-of-covid-19/recovery-and-rebuilding/whats-the-future-of-live-events-post-pandemic/

UN-Habitat. (2022). Climate change: Overview. UN-Habitat. Retrieved 25 March 2022 from https://unhabitat.org/topic/climate-change#:~:text=Urban%20areas%20are%20major%20contributors,vulnerable%20to%20climate%20change%20impacts

UNEP. (2014). Emerging issues for small island developing states. UNEP. Retrieved 2 February 2022 from https://sustainabledevelopment.un.org/content/documents/1693UNEP.pdf

UNFCCC. (2015). *The Paris Agreement.* UNFCCC. Retrieved 2 February 2022 from https://unfccc.int/process-and-meetings/the-paris-agreement/the

-paris-agreement#:~:text=The%20Paris%20Agreement%20is%20a,compared %20to%20pre%2Dindustrial%20levels

UNFCCC. (2021, 17 September). Nationally determined contributions under the Paris Agreement: Synthesis report by the secretariat. Conference of the Parties serving as the meeting of the parties to the Paris Agreement, Glasgow. Retrieved 1 February 2022 from https://unfccc.int/sites/default/files/resource/ cma2021_08_adv_1.pdf

UNICEF. (2021, 15 September). Schools still closed for nearly 77 million students 18 months into pandemic. UNICEF. Retrieved 24 March 2022 from www.unicef.org/press-releases/schools-still-closed-nearly-77-million -students-18-months-pandemic-unicef

United Nations Development Programme. (2015). *Sustainable Development Goals*. UNDP. Retrieved 1 February 2022 from www.undp.org/content/dam/ undp/library/corporate/brochure/SDGs_Booklet_Web_En.pdf

United Nations Environment Programme. (2021, 16 September). COVID-19 caused only a temporary reduction in carbon emissions – UN report. United Nations Environment Programme. Retrieved 25 September 2021 from www .unep.org/news-and-stories/press-release/covid-19-caused-only-temporary -reduction-carbon-emissions-un-report

United Nations Framework Convention on Climate Change. (2021). Glasgow Climate Pact. Decision/CP.26. UNFCCC. Retrieved 1 February 2022 from https://unfccc.int/sites/default/files/resource/cop26_auv_2f_cover_decision.pdf

van Zoonen, L. (2016). Privacy concerns in smart cities. *Government Information Quarterly*, *33*(3), 472–480.

Walton, D., & Aalst, M. v. (2020, September). Climate-related extreme weather events and COVID-19: A first look at the number of people affected by inter-secting disasters. IFRC. Retrieved 10 August 2021 from https://media.ifrc .org/ifrc/wp-content/uploads/2020/09/Extreme-weather-events-and-COVID -19-V4.pdf

Weng, M., Ding, N., Li, J., Jin, X., Xiao, H., He, Z., & Su, S. (2019). The 15-minute walkable neighborhoods: Measurement, social inequalities and implications for building healthy communities in urban China. *Journal of Transport and Health*, *13*, 259–273.

West, D. M. (2018). *The Future of Work: Robots, AI, and Automation*. Brookings Institution Press.

World Bank. (2020, 8 June). The global economic outlook during the COVID-19 pandemic: A changed world. World Bank. Retrieved 22 April 2021 from www.worldbank.org/en/news/feature/2020/06/08/the-global-economic-outlook -during-the-covid-19-pandemic-a-changed-world

Yang, J., & Reuter, T. (2020, 16 March). 3 ways China is using drones to fight coro-navirus. World Economic Forum. Retrieved 24 March 2022 from www.weforum .org/agenda/2020/03/three-ways-china-is-using-drones-to-fight-coronavirus/

Yigitcanlar, T., Kamruzzaman, M., Buys, L., Ioppolo, G., Sabatini-Marques, J., da Costa, E. M., & Yun, J. J. (2018). Understanding 'smart cities': Intertwining development drivers with desired outcomes in a multidimensional framework. *Cities*, *81*, 145–160.

Zheng, Y., Yang, S., & Cheng, H. (2019). An application framework of digital twin and its case study. *Journal of Ambient Intelligence and Humanized Computing*, *10*(3), 1141–1153.

Index